Setting Up a Food Drying Business

Setting Up a Food Drying Business

A Step-by-step Guide

Fabrice Thuillier
GERES

Translated by Maighread Holland

Practical
ACTION
PUBLISHING

Practical Action Publishing Ltd
27a Albert Street, Rugby, CV21 2SG, Warwickshire, UK
www.practicalactionpublishing.org

© GERES

English language translation © ITDG Publishing 2002
First published in 2002

The French version of this book is published by GERES and was written by Fabrice
Thuillier, agrofood engineer, GERES with the collaboration of Anne Claire
Outtier, engineer, GERES; Elisabeth Aillaud and Marie Ange Le Bloch, typing; and
Nicolas Sainpy and Vahram Varjabedian, illustrations.

English translation by Maighread Holland

ISBN 978 1 85339 498 0
ISBN Library Ebook: 9781780441344

A catalogue record for this book is available from the British Library.

Maighread Holland has asserted her right under the Copyright Designs and Patents
Act 1988 to be identified as translator of this work.

Since 1974, Practical Action Publishing has published and disseminated books and
information in support of international development work throughout the world.
Practical Action Publishing is a trading name of Practical Action Publishing Ltd
(Company Reg. No. 1159018), the wholly owned publishing company of Practical
Action. Practical Action Publishing trades only in support of its parent charity
objectives and any profits are covenanted back to Practical Action (Charity Reg. No.
247257, Group VAT Registration No. 880 9924 76).

Typeset by Dorwyn Ltd, Rowlands Castle, Hants, UK

Contents

Preface

In countries of the South, food drying is an ancient activity, still practised today in town and country. The process is used for many products: cereals, root vegetables, meat products, green vegetables and seasonings.

The expansion of vegetable farms and the development of urban markets have opened up new perspectives for improving traditional drying. By selling dried products, you can increase the value of surplus seasonal produce and replace fresh produce when it is in short supply.

Drying offers a new opportunity in the small-scale food production sector, where there are increasing numbers of small businesses. For several years non-governmental organizations (NGOs) and research organizations have been trying to provide technical responses to requests from entrepreneurs. However, the success of a drying business does not just depend on mastering an appropriate technology – project planning, business management and commercial approach are equally important.

GERES has been working in the field of food drying for more than 10 years. This step-by-step guide aims to help those setting up projects and provide them with essential information. This book is also aimed at supporting bodies (donors, NGOs, research and development organizations) and interested groups that wish to understand the sector and contribute to its development.

This guide has four chapters, which are separate but complementary.

- The first chapter defines the drying units discussed in the book and the general context when creating a drying business.
- The second chapter is a methodological account of the procedure for setting up a project, and covers market identification, technical choices, quality processes and economic analysis.
- The third chapter covers the different markets for small-scale drying businesses, including technical notes on the main dryers available.

- The fourth provides possible avenues for improving the profitability of drying and adapting dryers to suit the needs of businesses.

GERES wishes to thank the many people who contributed to the production of this book. The editing and publishing were made possible with the help of the European Commission's Directorate General for Development.

Acknowledgements

This book was written by Fabrice Thuillier, food production engineer, GERES, with the help of Anne Claire Outtier, research engineer, GERES. Thanks go to Elisabeth Aillaud and Marie Ange le Bloch for the presentation, and Nicolas Sainpy and Vahram Varjabedian for the illustrations.

Reading panel: Alain Guinebault, GERES; Nobert Monkam and Jean Calvin Dongmo, Agro-Pme (Cameroon); Jean Michel Méot, Cirad (France); Alain Traoré, ABAC-CAA (Burkina Faso).

Rewriting and making up: Imédia Dakar.

Chapter 1

The entrepreneur and the project

1.1 Small-scale food-drying units

Food drying varies greatly in terms of the scale of production and the products processed. It takes place at both domestic and commercial levels, and involves basic food products such as root crops or market garden produce such as vegetables and seasonings.

Here we consider small-scale food-drying units, focusing on production processes with the following characteristics.

- They are carried out within a private business (whether or not it is legally recognized) and managed on an individual or group basis, where their general characteristics can be clearly recognized (location, financial backing, human resources, means of production).
- Drying is the main activity of the unit, even if the product or products are the subject of pre- or post-processing.
- The aim is to generate income by adding value to locally produced food products.
- The expertise is small-scale in nature, regardless of the level of production, size of business or market targeted.

So, for example, this term could represent a group of tomato producers established in the area around a town who dry any produce they have not sold and sell it in town; or a woman trader who exports ground, dried cassava.

1.2 Significant development potential

The 1990s saw the emergence of the African small business, at the same time as large, partly state-owned projects were gradually disappearing. The great increase in population, and the need to feed towns with ever-increasing populations, brought about huge growth in the

1

Figure 1.1 Two examples of products from small-scale food-drying units.

food-processing industry, where official and informal sectors work side-by-side. Today small-scale food production represents a very fast-growing sector.

In this context, drying of food products has significant development potential. New income-generating activities can be introduced that respond simultaneously to the need to preserve harvests, particularly for perishable foodstuffs, and to a new demand on the part of urban consumers for stable, ready-to-use products. Fulfilling this potential requires the development of techniques and a capacity for innovation in order to adapt to these new consumers and offer them new products.

However, the potential for using drying to add value to food products should not obscure the difficulties involved in breaking into markets. A dried product will be sold at a higher price, even if it is of better quality than the same product offered in its traditional, unprocessed state. If a new product is involved, it will need to be advertised and to become known and appreciated. Local urban markets will, therefore, no doubt be a limited market, especially in view of the weak buying power of consumers.

The solution to developing a market in dried products is perhaps to be found outside the region. The freeing-up of exchange mechanisms favours sub-regional trade and exporting to industrialized countries. These products have the advantage of travelling easily. However, this international dimension presents difficulties for small-scale production units: this approach requires a clear strategy and considerable funds, and involves some risk.

1.3 An essential partnership

Drying activities may be very varied – and the promoters of drying units may also come from very different backgrounds: male or female, farmer or trader, from an urban or rural environment. Above all, an entrepreneur is needed with the skills to gradually build up the project.

A single individual can rarely develop such a project alone. An entrepreneur will need to choose partners who are able to guide and support him or her in the different aspects that should be taken into account: technical, financial, commercial, accounting, and management.

One of the risks of partnership is that it is often fleeting in nature. Sometimes there is an imbalance between the rhythm of a project's life and the support that some of its partners can provide. Technical support programmes often have a short lifespan, and their aims are more quantitative than qualitative. Local and international organizations that provide loans have methods of financing which do not always fit in with the needs of developing projects.

GERES has been working for several years to develop a partnership that it describes as 'accompanying', for several reasons:

- it fits in with the lifespan of the project, adapting to the project's development and inspiring mutual confidence between the different participants
- it involves a maximum of components – technical engineering, commercial surveys and economic assessment – in a way that maintains the coherence of the project
- it mobilizes local resources for this purpose.

This personalized partnership makes it possible to bring together people with common aims, to solve problems together. However, it is difficult to find the financial resources necessary for such a step: the small number of 'accompanied' groups causes a certain failure rate, and sponsors often give greater weight to quantitative criteria, to the detriment of quality aspects.

The route of the entrepreneur

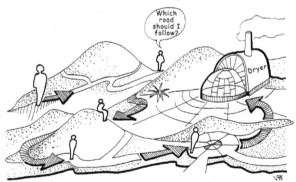

Figure 1.2 The route of the entrepreneur.

Chapter 2

Constructing the project

2.1 Taking different aspects into account

Creating a drying business is not just a matter of identifying a raw material and investing in a dryer. Several interacting factors have to be taken into account to form a coherent whole.

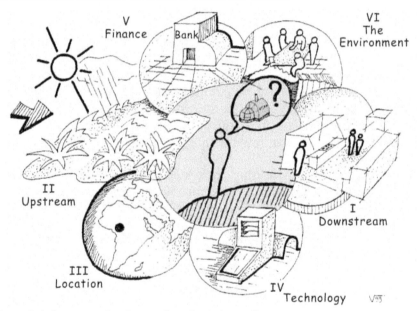

Figure 2.1 Components of a small-scale drying unit.

- Downstream linkages refer to the movements of finished products from when they leave the unit until they reach the consumer – including market, competition, market position, distribution network, etc.
- Upstream linkages refer to the raw materials and inputs necessary for production, and particularly include issues of availability, location, suppliers, forwarding merchandise, and marketing brief or briefs.

4

- Geographical location covers local climatic conditions, problems of accessibility and transport (with respect to both raw materials, equipment and energy) air quality (possible pollution), and servicing terrain.
- Technology system refers to the whole processing chain – the progression of the various sections of work. Apart from drying itself, a unit will be productive only if the associated operations for preparing and then packaging products are properly organized.
- The financial aspect includes all those who contribute to the funding of the drying unit apart from the individual entrepreneur – partners or shareholders, banks, financing organizations, etc.
- Environment signifies the various people with whom the unit has links, whether formalized or not – technical partners, competing businesses, administrative services, professional links, etc. This term also includes the statutory or normative system that the entrepreneur must take into account.

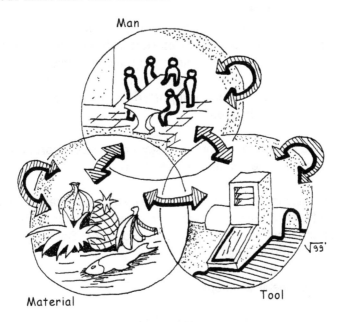

Figure 2.2 The concept of a technical system.

The size of each operation (machine capacity as opposed to workforce productivity) should allow for coherent linking of the various stages. Within a single stage, the tool of production has to be appropriate to the product formulation (preserving the colour) and the technical level (qualification of staff).

5

2.2 Approaching the future market

The market study is the most important stage in designing a project. It is no use identifying a raw material and deciding on a processing method if there are no existing markets. So it is necessary to start with consumers' needs and eating styles, and their possible development, with a view to producing processed products that are well adapted to customers' requirements.

Studying market potential

Before launching a product, the entrepreneur should study its market potential. Steps should be taken in advance to identify existing demand, study competing supply, and try to find customers who are not consumers but who might potentially be interested. Figure 2.3 illustrates these steps.

The existing total market

Identifying the total market for a product provides general figures for a particular area, including total market volume (in quantities and value); volume of imports by country; apparent consumption (quantity of product consumed in a given area in relation to volumes produced in the area + volumes imported – volumes exported); eating habits; use of the product, etc.

Example: dried mango for export to western countries (apart from second quality consumed in the country, 2–5 per cent)

- European markets: shared among specialized grocery trade (5 per cent), organic network (25 per cent), fair trade network (70 per cent).
- Origin of products: Asia (quantity not provided), South America (30 tonnes), Africa (60 tonnes).
- Consumption: food products, snacks.
- Points of sale: specialist shops.

The study of the overall market within a country should enable points of sale to be identified. This information is often difficult to find – research may be carried out via libraries and information centres, trade unions and associations for the business concerned, ministries (industry, fishing, etc.), statistical organizations, or aid and development organizations. A more detailed study can enable the large distribution circuits and development trends to be identified.

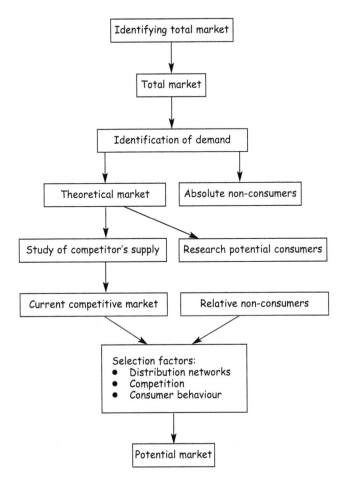

Figure 2.3 Procedures for identifying market potential.

Example: distribution circuits for French organic product market (1994)

Market share:
Health food stores	33.3%
Large and medium-sized supermarkets	23.3%
Export	16.7%
Farm shops	13.3%
Cooperatives	6.7%
Organic markets	6.7%

Theoretical market
The second stage involves identifying demand, making it possible to eliminate the category of 'definite non-consumers' – those people

who will not be interested in the product. This task can be obvious (as in the case of baby formula, for example, which involves only newborn babies), or it may have to be carried out with the help of surveys among retailers and consumers. These surveys (Appendix 1) enable a better understanding of consumers' needs and expectations.

Market potential
Moving on from the theoretical market, the definition of market potential takes into account the current competitive market and the relative non-consumer market.

- The competitive market can be evaluated by studying similar products offered in sales outlets: wrapping and packaging, selling price, quantities sold, names and addresses of competitors. This study can be taken further through surveys of sellers, in order to establish competitors' make-up (nature, number, significance and weight in the market); supply (characteristics and supply of products, brands, types of packaging); and policy (prices, products, means of distribution, promotion and sales forces).
- Study of the market of relative non-consumers is less easy: it involves both sifting through customers who might possibly be interested in a product, and discovering their reasons for not being satisfied with what is currently on offer. This study can be carried out at the same time as a theoretical market survey, the questionnaire being broadened to all customers in the area of distribution (Appendix 2).

The potential market can be segmented – divided into groups of consumers – according to criteria including socio-professional group, sex, age bracket, etc. Thus a particular segment may be chosen as the preferred target for the product.

Positioning the product

The next stage involves giving the product a specific position in the minds of consumers, so that they can distinguish it clearly from the competition. This presupposes that the needs to be met by the product have been determined in advance, and customers' appreciation of competing products and their evaluation criteria have been analysed. Positioning the product makes it stand out from the competition, and attract customers.

Key points in marketing strategy
Once the target and positioning have been defined, marketing strategy, or marketing policy, will be built around four areas.

- Product policy: the product must be designed on the basis of customers' expectations, taking into account not just its functional use, but also the symbolic image that the customer projects. So, the inherent characteristics (formula, design, standards) of the product must be fixed, and the packaging designed.
- Pricing policy: price plays an important role in customers' buying behaviour and appreciation of competing products. Fixing prices depends on several factors – the manufacturing cost of the finished product; competitors' prices; the aims pursued (a high-price policy projects an image of quality or luxury, a low-price policy enables you to attract most customers); and product value on the market (value includes originality, quality, whether there is plentiful supply of alternative products).
- Distribution strategy: the impact of distribution on sales of a product is very important. The choice of distribution network for the product should be able to guarantee – the most efficient access possible to the intended target; high-quality shop presentation; and regular flow in terms of quantity and frequency.
- Promoting products: promotion involves activities carried out to develop awareness of the product and make it appealing. The means of promotion can be very varied – direct sales to customers (door-to-door sales, taking part in exhibitions or trade fairs); advertising (publishing catalogues, broadcasting radio messages); sales promotion (sending samples, speculating on a policy of low pricing); and action on the product itself (commercial name, logo, packaging).

Example: product policy – the case of dried 'organic' mango
Intrinsic characteristics:

- presented in slices, orange colour
  natural sweet taste
  supple texture

Packaging:

- silver paper with a label that presents the product as an energy provider tasty snack for sports use – to target customers
- polyethylene window (2 × 2 cm) to allow visual appreciation of the product

- individual portion (net weight 50 g) for easy use
- consumer information about the technical characteristics of the product: list of ingredients, nutritional value, use-by date
- supplier details to ensure customer loyalty later.

Approaching international markets

Exporting products will affect drying businesses that already have production and sales experience in their national market and wish to enhance the value of their expertise in order to develop their market. Before considering such a policy, it is necessary to have a good command of the internal market, to have sound finances, and to provide products of recognized quality.

The process of approaching international markets has several stages.

Information gathering
First, this means gathering general information on the countries involved, including population, GNP, level of debt, inflation rate, regulations, commercial agreements in force, socio-cultural attitudes, image of exporting country within the target market. This information can be provided by embassies, specialist organizations' publications [in France, Centre Français du Commerce Extérieur; the MOCI (*Le Moniteur du Commerce International*) guide; Atlaxco] or from international organizations' web-based databases (the World Bank, the United Nations, UN Conference on Trade and Development, UN Industrial Development Organization, the Food and Agriculture Organization of the UN, International Monetary Fund, European Bank for Reconstruction and Development, Organisation for Economic Cooperation and Development and the UN International Trade Centre, among others).

More specifically, information is needed on the dried-products sector in the target country: this information includes size of target market; statutory constraints (health, technical, customs formalities); and existing marketing and promotional methods (trade fairs, exhibitions). This knowledge makes it possible to work out all the costs of approaching the market. These costs also include customs duties, and costs relating to distribution and promotion policy when establishing a market position.

Export diagnosis
Export diagnosis is used to evaluate the strengths and weaknesses of a business. It involves analysing the business's strong points and its

limitations by means of a series of questions relating to technical, commercial, financial and human aspects. A sample questionnaire is shown in Appendix 3. The list of questions given is not exhaustive – the more there are, the more finely tuned the diagnosis will be.

The weak points exposed by this diagnosis should be given particular attention; at the same time, measures could be implemented in order to produce additional strengths.

Market research
Investigating international situations usually means looking for intermediaries who can distribute the products in the new market. Research is all the more effective when carried out inside the country through direct contacts. Importers are significant players in the success of the operation, because they provide the interface between supply and demand (Chapter 3).

An exploratory mission should be well prepared. The visit should be organized to comply with the laws and regulations of the host country. Care should be taken over building relationships: for example, information brochures could be prepared on the business and the product or products.

Seeking new business is conducted through interviews. The presentation interview involves presenting the business to the representative, providing them with brochures and samples, and describing potential agents and/or partners in the operation – the aim is to make a good impression and offer a serious, high-quality image. The negotiation meeting is a further stage, involving precise identification of a partner (legal status, entry on trade registers), defining their products, indicating a selling price, defining the precise place of operation, specifying their preferences in terms of method of payment, defining the validity period of the offer, obtaining practical information about local regulations and customs terms and conditions, and reaching agreement about settling disputes.

On returning from the visit, take every opportunity to consolidate the links developed in the course of the visit – several rules should be followed.

• Remain in contact with the partners you have met, contacting them again quickly and sending them items agreed at the meeting.
• Reply rapidly to faxes from partners.
• Always send samples of the highest quality.
• Make changes in conjunction with the partner, and keep them informed as to any possible changes to be made within the company (technical or administrative).

- Prepare the terms of contractual agreement – this means choose an 'incoterm' (see Appendix 4) and payment arrangements (Appendix 5).

2.3 Choosing a dryer

Only when an entrepreneur has understood the future market sufficiently clearly can he or she begin to define the production equipment. Sources of technical information on dryers are given in the Bibliography; here we give just a few methodology indicators to make these choices easier.

Acquiring a minimum of technical knowledge

As drying is often a traditional operation, people tend to think it is simple. However, the drying process involves many aspects (temperature, humidity, water content, air velocity), fluctuations of which have a determining influence on the quality of the finished product. The ability to control these aspects determines the success of drying operations. In the past, approaches to developing dryers have failed to take account of this complexity.

- Sometimes the approach is technical in nature, with insufficient emphasis on the entrepreneurial context when developing tools adapted to specific needs. Many prototypes are 'stillborn' within the confines of research centres.
- Sometimes the approach is simplistic, taking no account of a demanding specification when responding to the needs of the business and the market. How many 'appropriate' dryers are no longer appropriate once installed on the production site?
- Before choosing a machine, the entrepreneur needs to understand the principles involved in the whole process of water removal, in order to assess the suitability of supplies provided by dealers. Hot air drying or dehumidifying, sucking or blowing, natural or assisted convection, thermal heating power, ventilation flow, and energy output are all ideas or measurements with which they need to be familiar.

Entrepreneurs should also be (and keep) well informed about the different equipment available at the local or international level, by consulting catalogues from different suppliers and technical information sheets about drying machinery.

The size of dryer in relation to requirements also requires knowledge of physical measurements (water content, dry material content, water loss) in order to define evaporation capacity – the amount of moisture that has to be removed from the product in a given period of time. The Bibliography provides references to works on this subject.

Preparing, opening and sorting tenders

Drawing up specifications

The specification gives clear information on all the essential characteristics of the equipment. These characteristics can be divided into five categories.

- Characteristics required for finished product: involves technical information about the product including its shape, size and composition (moisture content, additives) which influence the decision as to which technical model to choose.
- Characteristics of the product upon entry: apart from general characteristics (shape, water content), this category particularly involves specific pre-treatments (blanching, citric acid, pickling) which give it new properties and affect the acceptable temperature when drying is complete.
- Production aim: specifications should include information about the aim of production in nominal capacity (that is, at optimum operation). This information is estimated according to the predicted medium-term demand, the sales rate of products, and the storage periods and conditions for raw materials. At this stage it is important to forecast the rate of increase in output of the business before arriving at the nominal capacity. Several assumptions about increase in output can be presented in order to forecast a certain modularity in the equipment.
- Energy supply: an energy process *par excellence*, drying requires energy in order to evaporate water from the product and transfer it into the air. This energy is needed in two forms: (i) thermal energy, which gives the air its capacity to evaporate [quantity of water absorbed per unit of air (g water/m^3 moist air; or g water/kg dry air)], and may be provided by gas, electricity, hydrocarbons, biomass or sun; and (ii) mechanical energy, necessary to work the fan for circulating air, and provided by electricity (grid, generator or renewable).
- Climate: temperature and humidity conditions, also wind and number of days of rain for the drying period/s, are factors that need to

be known for most drying equipment where performance is influenced by climate.

An example of a specification is given in Appendix 6.

Selecting offers for tender
The efforts made to draw up a project specification are justified only if the call for tender is as open as possible. Generally, there are only a few suppliers of drying equipment in African, Caribbean and Pacific (ACP) countries, but one should still try to obtain as many bids as possible to ensure that there is some competitiveness in the bids received.

Vetting the proposal for equipment
Analysing proposals from companies submitting tenders should take into account the following criteria.

- Compliance with the specification: does the proposal take account of all the assumptions listed? If not, can the equipment provider justify point-by-point the terms of his or her proposal? This analysis also relates to references about equipment (number of examples installed, locations, type of products processed).
- Technical level and training needs: the proposal should describe technical characteristics (components, control and monitoring instruments) and delivery terms (keys in hand, assembled in situ). These aspects affect skills and training needs required for setting up the drying plant and conducting operations. A minimum of local expertise during each of these operations is a factor in success.
- Setting-up and maintenance conditions: is the proposal compatible with the intended situation of the plant? Apart from the space that has to be available, account must be taken of energy needs, electric power to be installed, and the amount of heating energy to be stored over and above daily working requirements. Maintenance and servicing operations should be correctly defined and able to be carried out locally to avoid jeopardizing the smooth running of the operation.
- Managing upstream and downstream operations: each type of equipment imposes or favours a way of working (continuous or batch) which, according to production capacity, affects the organization of upstream and downstream stages – productivity, choice of additional equipment, manpower needs, etc.
- Financial and economic variables: the price of the tender is not the only thing to consider when choosing a proposal; several economic and financial factors must also be taken into account. These

14

include: (i) the lifespan of the machine and the frequency of replacing used parts – a more expensive machine often proves to be stronger and therefore better in the long term; (ii) drying cost, which depends particularly on energy costs (in kWh/kg of water evaporated, for example), is essential when calculating running costs; and (iii) additional costs such as training, technical assistance, and financial costs associated with loans.

Taking account of upstream and downstream operations

Upstream operations (preparing the product prior to drying) and downstream operations (shaping prior to marketing) are integral parts of the production system. It is essential that they fit in with the drying. Defining these operations, although less complex than the dryers, follows a similar method.

- Information is needed about operational procedures and additional equipment.
- Specifications must be drawn up on the basis of needs identified.
- Proposals from providers should be validated.

In the case of a drying unit, upstream and downstream operations are often not highly mechanized, so it is particularly important to define human resource needs according to the productivity of each operation and to calculate the space required for each one.

Example: lychees

The productivity of an employee peeling and removing stones from lychees (a small, soft fruit with a stone and peel, widely found in the Indian Ocean Islands and in South-East Asia) is 2–3 kg/h with a stone remover. The preparation of 300 kg of produce over a five-hour period requires the presence of 20–30 people.

Factors to be taken into account for upstream and downstream operations in drying work are presented in Table 2.1.

2.4 Quality considerations

The idea of 'quality' is increasingly evident in the food processing sector (see Appendix 7). As regards the product, quality can appear to be intrinsic (good or bad), but in fact is experienced differently according to customer perceptions. It corresponds both to expressed need,

Table 2.1 Upstream and downstream operations

Operation	Function	Factors to consider
Upstream operations and preparation for drying		
Arrival and sorting	Reject any material that does not meet requirements for drying	• Keeping notebook of loads • Purchasing responsibility • Reception area • Criteria and method for judgement
Storage	Keeping the product fresh	• Maximum storage time • Natural or artificial (refrigerated) conditions • Area and structure of storage • Planning and management of store
Washing	Cleaning the product before drying	• Manual operation: labour estimate, quality of water for washing, tank • Draining waste water
Preparation	Prepare the product so that it is in the required shape after drying	• Manual operation • Skill required • Task organization • Tools and utensils • Trained supervisor • Pre-treatment • Use of by-products
Downstream operations and product preservation		
Removal and sorting	Distinguish dried products by grade	• Manual operation • Quality of the tray base for removal • Specification by grade • Evaluate the level of rejects
Packaging	Packaging the product for preservation and sale	• Mechanized operation – manual or automatic heat sealing • Choice of container – preservation conditions and marketing criteria • Restrict the operation time
Downstream storage	Preserve the finished product for sale	• Know the ambient conditions • Protect stocks from natural external damage • Method of monitoring stock and of management (inventory) • Controlling access to the stocks

such as taste and texture, and implicit need, such as quality of hygiene, ease of chewing and digestion.

The idea of quality does not end with the product, but also relates to all the services within the chain that starts with the producer and ends with the consumer. In industrialized countries, customer demands, regulation and increased competition have brought ever more demanding standards in quality criteria. This tendency is increasingly found in countries of the South.

Managing quality also means setting up methods and systems of preventative action to guarantee the best possible control in operating the business. It aims to provide the best production conditions possible, while inspiring confidence and guaranteeing transparency for all participants inside and outside the business.

The manager of a small-scale drying unit should take account of these ideas from the outset. They aim to guarantee the 'quality' of the business, particularly from an economic point of view (see Chapter 4).

Constructing the unit

Site location

The choice of site for the unit should be made after examining different possible solutions. Account should be taken of factors such as proximity to suppliers of raw materials, and means of transport (railways, roads, vehicles). As the raw material (fruit or vegetables) is fragile and perishes easily, the quality of access routes must be studied with care.

Factors associated with the external environment are also important, for example, the proximity of a field treated with pesticides; a chemical treatment factory; the presence of dusty air; or the surrounding humidity.

Plan of unit

When planning the unit, certain rules should be followed.

- Calculate the ideal dimensions for each component on the basis of the number of people required for each operation, the position of the machines and accessories, and the constraints arising from the movement of people and products.
- Respect the principle of forward movement for food products.
- Divide the unit of production into several zones (dry, moist and buffer zones) allowing for necessary separations between them.

17

- Avoid cross-contamination because of movement of staff and product flow.
- Consider staff conveniences: cloakrooms, toilets.

A typical plan is given in Appendix 8.

Construction materials
The floors and walls must have the following.

- Smooth walls that can be cleaned with jets of water.
- Adequate drains (soakaway outlets).
- As few sharp angles as possible.
- A slight incline to help get rid of used water.
- Protection of openings against insects.

Organizing production and managing human resources

Organization prior to production
Before investing in the unit, it is advisable to carry out production trials in a research laboratory or in an organization with a technology room equipped with drying machines. The object of these trials is to get to know the precise characteristics of the raw materials (degree of ripeness, varieties, conditions for optimum preservation) and to identify all aspects of managing production.

When trials for formulating the product are being carried out, it is advisable to adopt a risk-analysis approach – the Hazard Analysis–Critical Control Points (HACCP) method. This method requires that corrective and preventative measures should be put in place at every stage in the process, in order to prevent the production of a defective batch from putting consumer safety at risk. The HACCP method is an integral part of a quality assurance system: applying it will be all the more profitable with the assistance of a firm of quality assurance consultants.

In concrete terms, preventative action takes the following form.

- Technical notes on formulation for each operation (quantity going in, length of operation, technical aspects such as temperature and humidity relative to the air).
- Plans for cleaning and maintenance.

Corrective actions propose stopgap measures for technical failures and offer alternatives to subsequent use of the defective batch.

Formulation trials should rely as much as possible on tasting trials: the process is then regulated in accordance with the results of sensory analysis.

Afterwards, introduction to commercial production is carried out with full equipment to full scale; maximum account should be taken of the technical aspects defined in the formulation phase.

Management of human resources

Effective management of human resources is an essential element when setting up a quality assurance system. Apart from all the precautions that management has taken at technical level (formulation notes, cleaning plan, postal forms, etc.), it must be able to involve all staff in the quality control process: this is achieved through recruiting skills that match the planned work positions, and through motivating staff who should be aiming for common objectives.

This motivation of staff can be achieved by planning a range of measures.

• Placing staff in a psychologically attractive environment (salary in accordance with work tasks and skills, good working conditions, easy communication with other workers and management).
• Giving them a period of training in carrying out the tasks required, possibly ending with a period of induction in the workplace.
• Ensuring transparency – making available regular information about directions and changes occurring in the organization.

For staff who have direct contact with food products, there should be systematic raising of awareness about hygiene and cleanliness. This may be carried out by someone inside or outside the organization.

Only when these conditions have been met, and staff are aware of their responsibilities, will they be able to carry out their work with real involvement and notice possible problems affecting production. It is important to have a good flow of information from management to staff, and vice versa.

To avoid serious corrective action later, it is important to plan for an effective working environment at the project design stage, in the interests of efficiency and quality.

Managing quality assurance

We have seen that a 'quality' approach means ensuring a quality production process (internal organization and design of the food

Figure 2.4 The road to quality.

product), and also knowing how the production process fits in with the whole supply chain – both upstream of processing (producers and suppliers) and downstream (distributors and post-processing).

Quality management upstream
To achieve a quality product, it is necessary to work with suitable raw material. Once the raw material characteristics have been precisely defined, the processor has to locate suppliers who can meet these specifications. Suppliers can then be selected according to the quality/price and service they offer.

The contract between supplier and processor must refer to terms of sale, defining the characteristics and supply of raw material. It should specify action to be taken if one of the clauses of the terms of sale is not met. For example, if one lot of delivered raw material does not meet the specifications required, it must be returned to the supplier at their cost.

Example: clauses in terms of sale for the supply of raw material

- Characteristics of raw material (for a particular type): variety required, weights, ripeness, grade.
- Transport conditions: if the product should be certified organic, some transport conditions should be specified (transport restricted to this type of use and authorized by the certifying authority).

- Frequency and times of delivery: it is possible to plan supply on a yearly basis according to the agricultural production calendar. In general, supply schedules are fixed for each day of the week.
- Cost of raw material.

Quality management downstream of processing

Integrating production into a quality system also means ensuring food safety (hygiene) in the whole distribution system, right up to the consumer. That is why the sale of food products to intermediaries or to the distributor is covered by a contract.

Example: clauses in a specification

- Characteristics of the end product: nutrition content, guarantee of food safety (results of microbiological tests), storage conditions, and duration in terms of best before dates* (e.g. 12 months for dried fruit) or consume before dates† (e.g. 20 days for fresh eggs).
- Delivery lead times and schedules.
- Price of the end product.

* Best before date: for products that do not need a consumption date, gives an indication of the period over which the product will maintain all its qualities.
† Consume before date: depends on the standards for each category of processed products, date after which a food safety risk is possible.

The concept of food safety relates to the problem of food poisoning; for every consumer complaint it should be possible to follow the supply chain back and identify at which point it failed. This process of precise indentification of production fault (being able to identify the date and time, the place and the person responsible) is called traceability. Every production system needs to be organized, numbering production lots with specific markings (time, date and sample number).

The purpose of quality control is to achieve complete customer satisfaction. This will encourage customer loyalty, continued success for the product and maximum profit for the entrepreneur.

- Hygiene quality means an essentially healthy product (with no bacterial pathogens). This is extremely important: the best before date must appear on the label, guaranteeing hygiene quality.
- Nutritional quality means a food product in which the biochemical composition meets the nutritional needs by category for individuals. This can be included on the label for optional use.

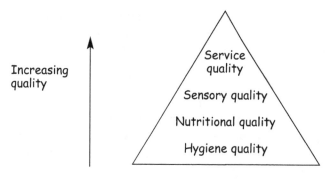

Figure 2.5 The quality pyramid.

- Sensory quality requires sensory standards from consumers. Taste trials enable the sensory qualities of a food product (sweet, bitter, sharp, crisp, smooth) to be identified.
- Service quality means, for example, ease of use of the product (practical packaging and accurate measurement an integral part of processing). Service quality is an optional element which can help in competing with other products on the market.

2.5 Conducting an economic analysis

Use of economic analysis

The final objective of economic evaluation is to study the cost-effectiveness of the business. It can be used for forecasting prior to launching (*ex ante* appraisal), and for appraising these forecasts once the project has begun (*ex post* evaluation).

Ex ante *appraisal*
Before investing in any staff or equipment, those starting a project must be aware of what is at stake economically and financially. The first thing that the economic evaluation covers is the feasibility of the business: 'taking into account the projected quantities envisaged, the intended means of production, the markets targeted, the projected investments and running costs, will my drying business be viable and will it be able to generate the expected profits?' The accounting mechanisms outlined below are intended to assist with this.

Apart from helping to assess feasibility, the *ex ante* evaluation is also a forecasting mechanism used to validate hypotheses, such as, for example, measuring the impact of a new investment or a drop in sales. Situations can be simulated by studying the sensitivity of certain factors.

The results of the economic evaluation become tools for promoting the project with partners and financial institutions, who generally require this information before considering whether to be associated with the project.

The ex post *evaluation*
The economic evaluation is also useful once production activities are under way. In addition to accounting, which helps to manage the flow of money, economic analysis sheds essential light on interpreting the economic reality of the business and helps to validate (or deny) the *ex ante* hypotheses.

The economic evaluation serves to interpret the results of the business (percentage of added value on turnover, distribution of fixed or variable costs, etc.), and thus becomes a means of making choices and retaining priorities, such as reducing a particular cost item, or fixing a maximum price for purchasing raw materials.

Accounting mechanisms

The economic evaluation makes it possible to put into figures the present or future operation of the business. This process uses reference tools (see Bibliography): the estimated operating account and cash-flow forecast, receipts and payments. They enable the company to organize economic and financial data in order of importance.

At the outset, it is essential to fix the data on production quantities: quantities produced and sold, quantities of inputs, consumables, utilities, raw material, property, plant and human resources.

In an economic evaluation, the timescale always covers a number of years – the reality of a business is judged over several years. The period to be taken into account is not always easy to determine: three years is the minimum if one is to have some visibility; some prospective evaluations can exceed ten years.

Estimated operating account: a true reflection of the business
This mechanism synthesizes the economic data relating to business production, in this case drying, in the form given in Table 2.2 (see a detailed version in Appendix 9).

This table initially shows the products: they are made up of sales receipts (called turnover) and operating costs, divided into charges which are directly proportional to production and fixed charges.

At this stage, the balance sheet brings out the idea of value added: it represents profit made on raw material.

Table 2.2 Headings in an estimated operating account

Turnover

−

Operating costs

=

Added value

−

Tax and duties

−

Depreciation allowance

−

Financing charges

=

Pre-tax profit

−

Tax on profit

=

Net profit

+

Depreciation costs

=

Cashflow

Then there are three headings:

- local macroeconomic data: such as direct and indirect taxes and duties in force in the country
- depreciation allowance: represents the amount to be put aside each year so that equipment can be renewed when it is no longer useful
- financing costs: the total amount of interest for loans that the business may have taken out to finance its assets, investment, operating capital, etc.

Once these deductions have been made, the net result corresponds to the profit generated by the production work for each year covered by the exercise.

To complete this table and make the transition to the following one (Table 2.3), the cashflow is calculated by deducting the depreciation costs (which are not an expense but a provision) in order to calculate precisely the amount of finance resulting from the exercise which can be made available to guarantee its continuation.

Cashflow forecast receipts and payments: a true reflection of the business
The estimated operating account allows the project manager to appraise the economic viability of one business as opposed to

Table 2.3 Headings for cashflow forecast receipts and payments: a true reflection of the business

Receipts
Own capital
Loan
Cashflow
Subsidies
–
Payments
Investments
Working capital requirements
Repayment of loans
Dividends
Miscellaneous
=
Cash balance

another. The cashflow forecast of receipts and payments (Table 2.3) is the key business statistic which enables management of the resources necessary for running the business. A detailed table is given in Appendix 10.

The capital of the business is used in the first instance to invest in production capacity: equipment and infrastructure (primary applications of funds).

At the same time, funds must be raised for working capital requirements. Working capital should cover all expenses that the business will have to pay out before being paid by its customers. This can be a relatively long time where export markets are involved. This requirement is estimated for the year when it is at its highest, and varied according to requirements for each year.

Although they are absolutely essential, these requirements for working capital are often forgotten in calculations. It is necessary to produce the first products before one can sell them, without being able to collect payment for them. Even when the customer might pay for some purchases in advance, they will never pay for all of them.

The capital of the business may come from various sources.

- At the outset they include the promoter's and possible associates' private funds. These can be supplemented by bank loans. Any loans obtained give rise to financial costs which are included in the estimated operating account, and repayment of capital comes under the heading of 'assets' in the cashflow forecast receipts and payments.
- At the end of the first exercise, a positive cashflow will ensure that the business is self-financing i.e. it will not need short-term borrowings.

25

At the end of each exercise the difference between receipts and payments gives the cash balance. This is the final indicator of profitability. The accumulated balance over the various years should be positive, for fear of facing bankruptcy.

When there are several partners (or shareholders) in the business, the cash balance is considerably reduced by the dividends due to them.

Using results

Drawing up the tables given above makes it possible to see, within the given timescale, whether the business is profitable and financially sound. It can be useful to calculate some simple or even more complex ratios in order to obtain profitability indicators.

Simple operating indicators

Production cost (C_p)

$C_p = C_o/Q$

Where C_o = operating costs (see Appendix 9) and Q is the corresponding quantity produced.

Manufacturing cost of the product (C_m)

$C_m = (C_o + C_f)/Q$, where C_f represents the fixed costs.

Break-even point
This indicator is calculated:

- when the sales price (P_s) is fixed: the minimum quantity (Q_{min}) which needs to be produced in order for the revenue to cover the enterprise's costs
- when the quantity (Q) to be produced is fixed: the minimum sales price $(P_{s\,min})$.

When the value added (VA) is zero:

$VA = 0$ (1), then $R = C_r$ (2)

where R = revenue and C_r = running costs.

To find the minimum quantity (Q_{min}):

(1) becomes $P_s \times Q_{min} = (C_p \times Q_{min}) + C_f$, thus $Q_{min} = C_f/(P_s - C_p)$

To find $P_{s\,min}$:

(2) becomes $P_{s\,min} \times Q = (C_p \times Q) + C_f$, thus $P_{s\,min} = C_p + (C_f/Q)$

These last two indicators are significant because they are an important reference point: Q_{min} as the production target to be reached in the given period; $P_{s\ min}$ as the limiting value in any negotiation with a client when a quantity Q is being discussed.

Viability indicators
The above indicators provide accurate information about the activity. Other indicators help the calculation of the yield of the activity for a certain period of time.

Payback time (t_p)
This is the minimum amount of time for the accumulated investment costs to be covered by the accumulated self-financing of the enterprise over the years. Table 2.4 gives an example of a four-year payback.

Table 2.4 Example of a four-year payback

Year	0	1	2	3	4	5	6
Investment	+16			4			10
Cashflow		2	4	6	8	10	10
Balance	−16	+2	+4	+2	+8	10	0
Cumulative balance		−14	−10	−8	0	10	10

Return on investment (ROI)
This indicator allows a comparison between two projects using a cumulative ratio over a given period, with the cash balance as numerator and the required criterion as denominator.

Example 1: two projects with an investment of 16 monetary units (MU) after 6 years

Project A: 1 cash balance of 12 MU

Project B: 1 cash balance of 18 MU

$ROI_a = 12/16 = 0.75$

$ROI_b = 18/16 = 1.125$

Example 2: two projects with the same cumulative balance of 24 MU

Project A: capital invested: 4 MU

Project B: capital invested: 6 MU

Project A has a better return than project B

The notion of discounting

The limitations of the above ratios are due to the fact that they are evaluated without taking account of the development of macro-economic effects (inflation levels, monetary depreciation, etc. over time). That is why the calculations of viability need to use the notion of discounting. We limit ourselves here to the definition of discounting, see the Bibliography for further information on the subject.

The net present value (NPV) is the present value of a unit of capital C received in year N at a discount rate (i)

$$NPV = C / (1 + i)^n$$

Chapter 4 gives some concrete examples to illustrate the effects of different components of economic analysis.

Chapter 3

Markets and production equipment

The previous chapters give ideas for setting up food-processing businesses in general. Chapter 3 specifically covers drying operations, and deals with markets and production equipment.

3.1 Principal markets

Local and sub-regional markets

Vegetables, seasonings and other traditional dried products
Products dried in the traditional manner and consumed in ACP countries are leafy, green vegetables (brassicas), *gombo*, peppers, tomatoes and onions and their leaves; they are mainly cooked in a sauce to go with basic rice or mashed cassava dishes.

Selling these vegetables in dried form provides an opportunity to add value to seasonal surplus produce, and a way of replacing fresh produce in times of shortage.

The expansion of market gardening raises a problem in keeping produce fresh. The challenge this presents is increasingly important: for tomatoes, for example, post-harvest losses may be up to 50 per cent in some areas. Improving traditional drying offers a solution towards which development organizations have been moving for some years.

However, marketing products is not that easy, and inevitably depends on the interest consumers have in the product.

* In rural areas, traditional dried products are consumed by the whole population. They are home-produced on a domestic scale or bought on the markets. The market for products produced by a small-scale unit is very limited: customers' poor buying power does not permit them to buy more expensive products, even if they are of higher quality.

29

Gabou: onion from Niger

Pepper from Benin

Figure 3.1 Examples of traditional dried products.

- In urban areas, the market breaks down into segments according to income levels. The customers for locally produced dried products are made up of the middle classes, or those who are better-off. Products are sold on specialist distribution circuits (collectors, wholesalers, semi-wholesalers, retailers). Their quality is often better than that of traditional products and they may, for example, be wrapped in plastic packaging. These customers are very demanding. They buy dried products to vary their diet at certain times of year and they appreciate their quick and easy preparation. Higher-income earners are less aware of this type of product: they can buy imported fresh vegetables all year round, even though they are more expensive. They are more likely to be interested in new products (see below).
- Some products other than vegetables and seasonings are traditionally dried, and can find market opportunities thanks to the establishment of small-scale drying units. This includes country products which are relatively local, such as smoked meat or fish. For example, *killichi* (dried, flavoured meat from Nigeria) is today produced in small-scale units and marketed in the towns of Niamey, N'djamena or Ouagadougou for customers who wish to combine quality with authenticity.

New products

In addition to these traditional products, most dried products from small-scale units, sold on local markets are products that could be classed as 'new'. Often already sold in their fresh state, drying makes them into a new product. Today there are three main approaches.

30

- Products originally for export: these are discussed briefly here, and covered in more detail in the following chapter. Entrepreneurs well understand the importance of being known locally, of attracting customers who live close to production units and are fond of 'exotic' products, whether they are foreign or native. This is the case for tropical, dried fruits intended for western countries: they are increasingly consumed as snacks in the countries where they are produced, in the seasons when they are no longer available in the fresh state.

Figure 3.2 A new product – *Couscous made from fonio.*

- Dried products based on cereals or root crops. This is a market with huge potential, on condition that small-scale production units can perfect the necessary technical procedures (this problem is covered in Chapter 4). The growth of towns and changes in eating habits open up new horizons for ready-to-use, traditional dried products. In Dakar, for example, current demand is barely satisfied by more than 10 units, each with a capacity of 150 kg per day.
- Other products, such as instant preparations for making up traditional drinks, or even dried potatoes, can furnish opportunities for diversifying production. In Burkina Faso, for example, instant *bissap* (a powder based on dried hibiscus flowers and sugar) is increasingly popular.

So there are a huge variety of products that can be sold on local markets, and provide outlets for small-scale units prior to export.

The export market for dried fruits

Here we discuss exporting to western countries, dealing with the example of tropical fruits. From the early 1980s the fruit sector has seen greatly increased international cooperation, intended as a means of promoting exports from the countries of the South. Today this continues to be a potential outlet for small-scale drying businesses, but access to these markets is difficult. The countries of western Europe (Germany, Netherlands, UK, Belgium, Austria, Switzerland and France) are the main consumers of dried, tropical fruits.

Products and customers
Dried fruits for export can be divided into two categories.

- Products to be eaten: dried fruits sold as they are, or mixed into a cocktail ('exotic fruit mix'). They are eaten as nibbles (aperitifs), or as an energy food (sport), or (like dried, powdered coconut) in confectionery.
- Intermediary manufacturing products: dried fruits are incorporated into other food products and offered in desserts (yoghurt, ice cream) or for breakfast (muesli).

(a)   (b)

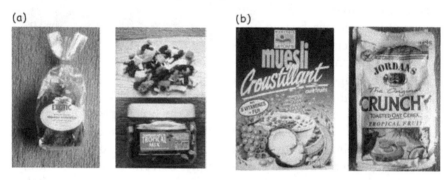

Figure 3.3 (a) Products for direct consumption; *(b)* manufactured products.

The European market in dry and dried fruit, spices and seasonings is dominated by specialist organizations. The importers are mainly importer–packagers, who buy the finished product in its original state and package it in accordance with the requirement of the manufacturer or the wholesaler. There are also intermediaries (brokers or merchants) who mainly provide contact between the supply from countries of the South and the demand from countries of the North. These importers will negotiate with the manager of the drying business who wishes to export their products (see Appendix 11).

In Europe, buyers may belong to one of three categories.

- Manufacturers: looking for products for use in manufacturing. Their marketing brief is demanding (calibre, texture, colour of products) and depends on processing procedures. Their criteria for providing supplies are also very strict.

- Importer–packagers: interested in all types of products, they supply the distribution market (small, medium and large supermarkets) with conventional products. They are interested in large volumes at low prices.
- Specialist importers: some may be interested in 'tropical' products (see below). Their advantage lies in low levels of competition and high purchase price. But they are few in number, their specification is demanding, and it involves a sector which is sometimes short-lived. The importer is, above all, a trader who is looking for the best compromise between quality, price, travel time and regularity of supply. Trade-offs are made according to the law of supply and demand. Particular attention has to be paid to the questions of exclusivity rights (which can pose risks when becoming dependent on a single supplier) and conditions of payment (see Appendix 4).

Organic and fair-trade markets

The conventional market changes little – it remains limited to a few partnerships developed with Asia, and leaves little room for diversification of the origins of products.

The more recent 'organic' and 'fair-trade' markets, on the other hand, offer real opportunities and are open to the countries of the South, particularly Africa and South America.

The organic market in Europe

With a market share of more than 1 per cent of total turnover for food consumption in the countries of the North, such as Germany, Austria or Denmark, and a prospective market share of 3 per cent in the short term, the organic sector is growing rapidly. Dried tropical fruits are still marginal, but they benefit from a 'ratchet effect' by now appearing in supermarkets.

But what is an 'organic' product? In 1991–92, the European Union brought into force regulations allowing the certification of an organic product according to precise specifications: absence of chemical products for synthesis; growing techniques; positive list of authorized non-organic ingredients permitted which can be up to 5 per cent of the content of the product (e.g. citric acid E330). Each organic product carries the label 'OF' (organic farming), granted by an independent certifying authority which controls production. The certificate is renewable every year. The certifying authorities now have representatives outside the European Union, in Africa and Latin America, for example.

Inevitably, the cost of certification and organic practices mean that the cost of these products is higher than others. But a section of consumers is ready to pay more if it has a guarantee that the product is healthier.

Supplying organic, tropical dried fruits means small volumes, but may be diversified: examples include pineapples, bananas and mangoes in Latin America (Colombia, Dominican Republic) and Africa (South Africa, Cameroon, Togo, Guinea); and coconut in Asia (Sri Lanka, Philippines).

Example: organic dried fruits already on the market

Pineapple is the most widely found product, coming mainly from South Africa. It is presented in slices, and sold in 50–70 kg packages [cost, insurance and freight (CIF) wholesaler; see Appendix 4].

Banana, the supply of which is medium in scale. The price varies from €5–8/kg (CIF wholesaler).

Mango, which is still little known, has only recently been imported from Burkina Faso. Demand is increasing, but the proposed

Figure 3.4 An example of organic dried fruits – pineapple.

product is still of mediocre quality (taste and colour). Its price varies between €6 and 10/kg.

The selling prices to consumers in this example are high. However, account should be taken of freight costs and margins for the middle-man or men. The current price to the producer (ex-factory) for mango, for example, is close to €5/kg. The quantities that it is possible to sell depend on the number of businesses with which the agent/importer has links. In the case of dried mango, they are made up of between 3 and 10 tonnes per season, according to the size of the business.

Even if the organic market strengthens, it still remains uncertain for the dried tropical fruits sector. This is a new market in which the small-scale drying units have to prove themselves. In particular, they must improve the quality of their products and the regularity of production.

Conditions for certification sometimes reinforce the difficulties encountered.

- Applying for organic certification is sometimes difficult in the countries of the South, and can be subject to dispensations or equivalence. Some producers or agents can be tempted to break the rules in order to satisfy profitability aims, which presents a considerable risk of loss of confidence on the part of the consumer.
- Certification is not necessarily attributed to a producer (a drying unit, in this instance). An importer can obtain certification in his or her own name for different production sites: the latter then become wholly dependent on this customer, as only those who hold a certification number will have access to the market.

The fair-trade market in Europe

The idea of 'fair trade' rests on the acknowledgement that there is an imbalance in the international exchange rates and an asymmetry in economic power between the countries of the North (consumers) and the people of the South (suppliers of raw materials).

The conditions for access to this market include the following.

- A guarantee of a lucrative producer price, pre-financing for their campaign, and a medium-term contract.
- On the other hand, the constraints of a communal type of processing business, and of social redistribution of some of the profits from the business.

The 'fair trade' in food products particularly involves cash crops (coffee, cocoa), and exotic dried fruits form only a small part of it.

Example: fruits on the 'fair-trade' market

- Dried mango: 30–50 tonnes/year. Burkina Faso takes up the market. Selling price: €6/kg [free on board (FOB); see Appendix 4].
- Pineapple: new product, but supply is growing, coming from Cameroon and Benin. Selling price: €7/kg (FOB).

Figure 3.5 An example of fruit on the 'fair-trade' market – dried mango.

Despite the small volumes involved, supply has not stopped growing and diversifying for ten years or so, mainly from African countries.

The major problem with 'fair trade' is still confidence, with products being marketed which are often of mediocre quality (absence of production controls), and the intervention of players and agents who sometimes lack professionalism (especially at marketing level).

Demand is likely to increase, as the European public is increasingly aware of inequalities on a global scale. This growth will certainly bring about an improvement in discipline and professionalism in this sector. One of the keys to this development lies in reinforcing the influence of producers.

Other products

In these two markets, particularly in the organic market, other products apart from dried fruits may interest importers; some examples – although data relating to them are almost non-existent – are as follows.

- Aromatic plants and stimulants such as mint, pepper and ginger, which come from tropical areas. They are sought after because supply is still limited.
- Tomato, which is still traditionally produced (dried in the sun) in the countries of southern Europe: current European regulations, which are more exacting than previously, condemn this method of drying.

3.2 Equipment

Dryers

The range of dryers adapted to small-scale businesses is limited, and there are few suppliers. This situation is the result of the relatively short history of small-scale drying, and the dissemination of technology that was not always the most appropriate in the context of production and management.

The criteria that have guided the development of drying machines have long rested on the principles of 'appropriate technology': simple, robust, reproducible, inexpensive equipment was needed. If some of these criteria are still valid, there are at least two others, which are not necessarily compatible, needed in order to provide a quality product (dried fruit pulp or ready-to-eat product) with an acceptable production cost.

- Equipment for flexible usage, that is capable of being adapted to diversification of products (dried vegetables but also plants or root vegetables).

- Efficient equipment, that is able to meet the requirements of the sector and the segment of the market for which it is used.

In addition, there are no companies in the ACP countries that make dryers. The equipment that is available, developed by NGOs, scientific research centres, international organizations and inventors, still has to be perfected. A significant amount of work remains to be done in order to develop local skills that can offer specialist services.

However, the situation is evolving, both in the North and in the South. Some European manufacturers are 'tropicalizing' their dryers. The local support structures are improving their technical skills, and are taking more account of economic aspects.

Data sheets are presented here for several dryers that are adapted for small-scale units, divided into three categories:

- dryers manufactured locally
- dryers imported from industrialized countries
- semi-finished dryers.

The precise details of designers and distribution networks are given in Appendix 14.

Dryers designed and produced locally
We make a distinction here between two types of equipment.

- Solar dryers (data sheets A and B): these work on the principle of heating air through the greenhouse effect, where the solar absorber and the drying chamber are the same.
- Fossil fuel dryers (data sheets C and D): these are vertical, multi-drawer mesh dryers, where the air circulates from top to bottom by natural convection. The burner uses gas or wood.

Dryers imported from industrialized countries
We present just a single model here – the dehumidifying dryer. Because of its high-technology nature this dryer cannot be classified in the category of intermediate technology dryers.

Dryers made locally and intermediate dryers all employ the principle of hot air to extract moisture from the product to be dried. On the other hand, dehumidifying the air is a process which enables moisture content to be reduced while at the same time keeping the temperature low, at less than 40°C. This process brings into play a refrigerating type of circuit (evaporator/condenser), as found on air conditioners.

Data sheet A IBE–ATESTA banco dryer

I Technical principles
1 Drying method	Hot air drying through mesh flooring
2 Energy source	Solar greenhouse only
3 Draught	Natural convection
4 Type of dried products	Fruit and vegetables

II Construction characteristics
1 Tray surface	12 m²
2 Ground plan area	16 m²
3 Dimensions	L = 6.2 m, W = 2.5 m, H = 0.8 m; 1.8 m
4 Principles and materials	Brick foundation, framework and trays of wood; solar collector plastic
5 Technical level	Average to demanding for construction of housing and framework
6 Imported component	High quality plastic film (polyethylene)
7 Upkeep and maintenance	Frequent replacement of the collector (tearing, opaqueness)

III Techno-economic characteristics
1 Dryer load	100–120 kg
2 Drying period	2–5 days (depending on insolation and ambient relative humidity)
3 Drying season	Limited to the dry season
4 Weekly throughput	20–60 kg dried fruit per week
5 Heating power	Solar collector
6 Draught flow rate	Max. 150 m³/hour
7 Controlled parameter	None
8 Energy consumption cost	None
9 Capital cost	€750 in Burkina Faso (1987)
10 Longevity	10 years

IV Status of dissemination and manufacturing conditions
1 Designer	First version in 1980: FAO–IBE (Institut Burkinabé de l'énergie)
2 Dissemination channel	ATESTA
3 Number in use	Limited dissemination
4 Countries of use	Burkina Faso
5 Manufacturers	Local artisans

V Advantages
Protected drying
Local manufacture
Large load
Reasonable capital cost

Disadvantages and requirements
Fragile solar collector, upkeep cost
Operating duration reduced by the type of operation
Long drying cycle
No control of drying

Examples of IBE–ATESTA banco dryers in Burkina Faso.

Data sheet B ITA–CERER cabinet dryer

I Technical principles
1 Drying method	Hot air draught dryer
2 Energy source	100 per cent direct solar greenhouse
3 Draught	Natural convection
4 Type of dried products	Grains, fish

II Construction characteristics
1 Tray surface	10 m^2 (on two or three levels)
2 Ground plan area	10 m^2
3 Dimensions	–
4 Principles and materials	Wooden framework covered with plastic film
5 Technical level	Fairly simple, local artisan
6 Imported component	None
7 Upkeep and maintenance	Replacement of plastic film

III Techno-economic characteristics
1 Dryer load	80 kg (cereals)
2 Drying period	2–4 days (according to insolation and relative humidity)
3 Drying season	Dry season
4 Weekly throughput	50–100 kg dried product per week
5 Heating power	0
6 Draught flow rate	0
7 Controlled parameter	None
8 Energy consumption cost	0
9 Capital cost	€450–600 in Senegal (1993)
10 Longevity	Variable

IV Status of dissemination and manufacturing conditions
1 Designer	CERER (Senegal)
2 Dissemination channel	CERER, ITA, local manufacturers
3 Number in use	Several units
4 Countries of use	Senegal
5 Manufacturers	Local artisans

V Advantages	Disadvantages and requirements
Protection of products	Limited period of use and dependent on climate
	Slow
Local manufacture	Fragile collector, cost of upkeep
	Design improvement needed (air volume excessive)

The ITA–CERER cabinet dryer.

Data sheet C ATESTA double-cabinet dryer

I Technical principles
1 Drying method	Hot air draught across trays
2 Energy source	Direct heating by gas burner
3 Draught	Natural convection
4 Type of dried products	All sufficiently porous products with thin skin (fruits, vegetables); avoid flours and cereal products

II Construction characteristics
1 Tray surface area	14 m^2 (2 × 10 trays)
2 Ground plan area	3 m^2
3 Dimensions	L = 2.3 m, W = 1.1 m, H = 2.2 m
4 Principles and materials	Banco foundation
5 Technical level	Artisan, local carpenter
6 Imported component	None except for instrumentation: bimetal strip thermometer
7 Upkeep and maintenance	Trays, air-tightness (joints)

III Techno-economic characteristics
1 Dryer load	100 kg fruit
2 Drying period	24–36 hours
3 Drying season	All seasonal
4 Weekly throughput	80–100 kg dried product per week
5 Heating power	2 × 300 W
6 Draught flow rate	200 m^3/hour per cabinet (evaluation)
7 Control parameter	Inlet temperature
8 Energy consumption cost	2.5–3 kWh per kg moisture removed
9 Capital cost	€1350 in Burkina Faso (1998)
10 Longevity	7–10 years for the structure

IV Status of dissemination and manufacturing conditions
1 Designer	ATESTA
2 Dissemination channel	ATESTA (Ouagadougou)
3 Number in use	Over 50 in Africa
4 Countries of use	Burkina Faso, Benin, Senegal, Morocco
5 Manufacturers	Trained local artisans

V Advantages
Local manufacture
Affordable capital cost
Rapid drying
Satisfactory hygiene quality
of dried products

Disadvantages and requirements
Reliability of the ramp burner (local)
Consistency of drying
Supervision, upkeep of the trays
Energy efficiency
No air outlet adjustment
High drying cost

Left, ATESTA dryers in Ouagadougou; *right*, interior of a cabinet.

Data sheet D AGRICONGO wooden cabinet dryer

I Technical principles
1 Drying method	Removal by hot air across trays
2 Energy source	Wood fire
3 Draught	Natural convection
4 Type of dried products	Any products in thin layers, except dense flours

II Construction characteristics
1 Tray surface	5 m² (five trays)
2 Ground plan area	3 m²
3 Dimensions	3 m high
4 Principles and materials	Base refractory brick, sheet metal cabinet and chimney
5 Technical level	Local artisan
6 Imported component	None
7 Upkeep and maintenance	Trays, chimney sweeping

III Techno-economic characteristics
1 Dryer load	50 kg
2 Drying period	2–4 days (depending on season)
3 Drying season	All seasons
4 Weekly throughput	20–50 kg per week
5 Heating rating	Not determined
6 Draught flow rate	–
7 Controlled parameter	Flue damper flap
8 Energy consumption cost	Not determined
9 Capital cost	€225
10 Longevity	To be assessed

IV Status of dissemination and manufacturing conditions
1 Designer	AGRICONGO
2 Dissemination channel	AGRICONGO
3 Number in use	Experimental stage
4 Countries of use	Congo Brazzaville
5 Manufacturers	Local artisans

V Advantages	Disadvantages and requirements
Capital cost	Energy availability
Local manufacture	Drying operation
	Temperature control
	Tray changing, supervision

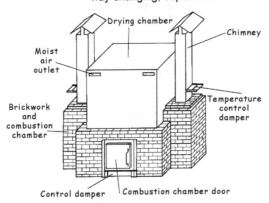

Figure 3.6 View of the AGRICONGO wood-fuelled dryer.

41

Data sheet E Dehumidifying dryer

I Technical principles
1 Drying method	Hot dry air over the products
2 Energy source	100 per cent electric from hot/cold batteries
3 Draught	Recycled forced draught
4 Type of dried products	Any products in thin layers

II Construction characteristics
1 Tray surface	2–8 trolleys
2 Ground plan area	–
3 Dimensions	–
4 Principles and materials	Fabricated: polyurethane panels, trolleys and trays stainless steel
5 Technical level	Specialized manufacturers
6 Imported component	Imported complete
7 Upkeep and maintenance	Maintenance contract

III Techno-economic characteristics
1 Dryer load	200–800 kg
2 Drying period	Less than 20 hours
3 Drying season	All seasons
4 Weekly throughput	With two trolleys: 300 kg dried fruit/week
5 Heating power	7–70 kW
6 Draught flow rate	2–5 kW
7 Controlled parameter	Temperature and relative humidity
8 Energy consumption cost	0.5–1 kWh per kg water removed
9 Capital cost	€375–1200 (FOB Europe)
10 Longevity	8–10 years

IV Status of dissemination and manufacturing conditions
1 Designer	European suppliers
2 Dissemination channel	Manufacturers
3 Number in use	Several units in Southern countries
4 Countries of use	Mainly European
5 Manufacturers	European suppliers

V Advantages	Disadvantages and requirements
Technical performance	100 per cent electric
Quality	Capital cost
	100 per cent imported technology
Process control	Requires a specific environment with standby generator.

Two dehumidifying ATIE dryers: batch type (left) and continuous type (right).

Intermediate dryers

Intermediate technology dryers are the result of analysing the limitations of the two types described above. They have two simultaneous aims: to meet the requirements of operators to whom imported equipment is not accessible and local supply is too limited; and to meet the requirements of the production specification (drying processes, operating cost, level of technical knowledge, etc.).

The intermediate technology models arose from the concept of producing a dryer that has already been tried out in a similar context and, as far as possible, should be locally manufactured. This requires the combination of local materials with imported equipment, and makes it essential to bring together a multidisciplinary team, made up of some who have knowledge of the technology and others who are responsible for local adaptation. This approach has several advantages.

- It offers a certain guarantee from a technical point of view, in that it proposes a tried-and-tested, efficient model, essential for the continuous running of small-scale drying units.
- It reduces investment cost to a minimum compared with the equivalent imported equipment – this cost is often a limiting factor when setting up a business.
- It favours local, appropriate technology – to mobilize local engineering is to gradually transfer knowledge of design, construction and maintenance.

Today the main limitation of this approach essentially lies in the cost of transferring knowledge (training, engineering), which is not an integral part of the investment cost and has to be financed from elsewhere.

The first three dryers presented here (data sheets F–H) are dryers designed by technology development organizations (Solagro, ITDG, Hohenheim) for specific situations and in a particular context. Plant dryers in the south of France (F), producer groups in Latin America (G) and North Africa (H) are distributed on demand, and are made to order with the support of the designing organizations. The Cartier dryer (I) is different because it comes from the French industrial sector of the 1960s. Equipment manufacturers continue to market it for drying businesses in Europe; the model shown here was first built in Africa (Burkina Faso) in 1995, and since then has been reproduced in Madagascar and the Ivory Coast.

Data sheet F Passiflore dryer

I Technical principles

1 Drying method	By hot air over trays
2 Energy source	Pre-heating 100 per cent solar or with complementary gas fuel
3 Draught	Forced draught
4 Type of dried products	Any products in thin or thick layers (except flours)

II Construction characteristics

1 Tray surface	10–36 m² models (in trays) + box option
2 Ground plan area	Collector roof 20–80 m²
	Drying chamber 1.5–6 m²
3 Dimensions	Built into the building
4 Principles and materials	Double roof: black steel plate or with glazing
5 Technical level	Local artisan, masonry
6 Imported component	Fan, filter, heater
7 Upkeep and maintenance	Trays, cleaning

III Techno-economic characteristics

1 Dryer load	100–350 kg (fruit) to 500 kg (plants)
2 Drying period	24–72 hours
3 Drying season	Solar: dry season; heater: any season
4 Weekly throughput	Depending on the model with heater from 50–500 kg/week
5 Heating power	Variable
6 Draught flow rate	300–3000 m³/hour
7 Controlled parameter	Inlet temperature, outlet air relative humidity (optional)
8 Energy consumption cost	Electricity: 0.3 kWh/dry kg
9 Capital cost	€2250–7500
10 Longevity	8–10 years

IV Status of dissemination and manufacturing conditions

1 Designer	SOLAGRO (France)
2 Dissemination channel	GERES, GEFOSAT
3 Number in use	7–8
4 Countries of use	France and one unit in Burkina Faso
5 Manufacturers	Local artisans and engineering by GERES–GEFOSAT

V Advantages
Rational use of energy
Use of the building roof
Flexibility of use

Disadvantages and requirements
Custom built
Capital cost

Passiflore dryer: view of the roof collector in Banfora, Burkina Faso.

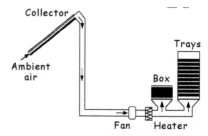

Figure 3.7 Schematic diagram of the passiflore dryer.

Data sheet G ITDG semi-continuous dryer

I Technical principles
1 Drying method	Hot air over trays, semi-continuous loading
2 Energy source	Oil- or gas-burning heater
3 Draught	Forced draught
4 Type of dried products	Sufficiently porous product to allow air flow

II Construction characteristics
1 Tray surface	12 m²
2 Ground plan area	1.7 m²
3 Dimensions	Height 2.2 m
4 Principles and materials	Heat exchanger with air conduit, wooden drying chamber
5 Technical level	Fairly complex, specific training needed
6 Imported component	At least heater, burner and fan
7 Upkeep and maintenance	Burners, trays

III Techno-economic characteristics
1 Dryer load	4.5 kg (plants)
2 Drying period	14 hours (duration for each tray)
3 Drying season	All season
4 Weekly throughput	4 kg/hour, thus 400 kg/week
5 Heating power	60 kW installed
6 Draught flow rate	2800 m³/hour
7 Controlled parameter	Inlet temperature
8 Energy consumption cost	Thermal: 2.5 kW per kg water removed Electric: 0.25 kWh per kg water removed
9 Capital cost	€4500 (FOB UK, 1993)
10 Longevity	Around 10 years

IV Status of dissemination and manufacturing conditions
1 Designer	ITDG UK
2 Dissemination channel	ITDG via local projects
3 Number in use	More than 10
4 Countries of use	Latin America, Asia
5 Manufacturers	ITDG and local partners

V Advantages	Disadvantages and requirements
Good productivity	Complex construction
Dissemination structure	Capital cost
Good energy efficiency	Work organization constraint
Low operating requirements	

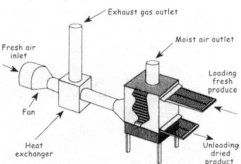

Figure 3.8 Functional diagram of the ITDG semi-continuous dryer.

ITDG semi-continuous dryer.

Data sheet H Hohenheim tunnel dryer

I Technical principles
1 Drying method	Hot air over the products
2 Energy source	Pre-heating solar collector
3 Draught	Forced draught
4 Type of dried products	Various

II Construction characteristics
1 Tray surface	20 m²
2 Ground plan area	36 m² (16 m² of collector)
3 Dimensions	L = 18 m, W = 2 m, H = 1.8 m
4 Principles and materials	Build in brick; trays metal frame with metal mesh grill, plastic covering
5 Technical level	
6 Imported component	Fan plus photovoltaic panels
7 Upkeep and maintenance	Upkeep and replacement of the plastic sheet

III Techno-economic characteristics
1 Dryer load	200 kg fruit
2 Drying period	2–3 days
3 Drying season	Dry season
4 Weekly throughput	80–100 kg dried per week
5 Heating power	Around 45 kWh/day solar energy
6 Forced draught power	20–70 W for 400–1200 m³/hour
7 Controlled parameter	Temperature as a function of air flow
8 Energy consumption cost	0.3 kWh per kg water removed
9 Capital cost	€1600 in Germany
10 Longevity	Building: 7 years; covering: 2 years

IV Status of dissemination and manufacturing conditions
1 Designer	University of Hohenheim
2 Dissemination channel	Innotech (Germany)
3 Number in use	Several tens
4 Countries of use	Asia, Anglophone Africa, Latin America
5 Manufacturers	Farmers' cooperatives

V Advantages
Effective drying thanks to its forced cover
Large capacity
(Mostly) local construction
Low energy consumption

Disadvantages and requirements
Dry-season operation
Replacement of the collector
Tray changing
Efficiency of the dryer
Quality of the dried products

(a)

(b)

(a) Putting ginger and chilli peppers on trays; *(b)* drying coffee in Indonesia.

Data sheet I Cartier tunnel dryer

I Technical principles
1 Drying method	Hot air over the products
2 Energy source	Gas or oil burner (with heat exchanger)
3 Draught	Forced recycled convection
4 Type of dried products	Any product (including flours) in thin layers

II Construction characteristics
1 Tray surface are	Two to six trolleys: 45–135 m²
2 Ground plan area	10 m² (for two trolleys)
3 Dimensions	Height 2.5 m
4 Principles and materials	Brick construction, metal trolleys, wooden trays
5 Technical level	Fairly complex
6 Imported component	Heater and fan plus tray mesh
7 Upkeep and maintenance	Greasing the fan, trays

III Techno-economic characteristics
1 Dryer load	Two trolleys: 320 kg; six trolleys: 960 kg
2 Drying period	18–20 hours
3 Drying season	All seasons
4 Weekly throughput	For two trolleys: 300–380 kg dried/week
5 Heating source	For two trolleys: 40 kW
6 Draught flow rate	8000 m³/hour (consuming 0.9 kW)
7 Controlled parameter	Inlet temperature, air relative humidity at exit
8 Energy consumption cost	1.5 kWh per kg water removed
9 Capital cost	€10,000 (in Burkina Faso), of which 50 per cent imported (1995)
10 Longevity	8–10 years

IV Status of dissemination and manufacturing conditions
1 Designer	Interprofessional Prune Bureau (M. Cartier), France
2 Dissemination channel	GERES
3 Number in use	Several tens of units in France, two in Africa
4 Countries of use	Burkina Faso, Madagascar
5 Manufacturers	Selected local artisans and GERES engineers

V Advantages
Advantages	**Disadvantages and requirements**
Technical performance (due to drying speed, energy efficiency)	High capital cost
Control	Construction engineering
Flexibility of use	Know-how requirement
Low operating requirements	Management capacity

(a) (b) (c)

(a) Exploded view of tunnel dryer; *(b)* hygrometry measurement at tunnel outlet; *(c)* fan.

Essential, additional equipment

We have seen that drying should be an integral part of a production process that includes upstream and downstream operations, which themselves require customized equipment. The choice of material is also limited, and depends above all on the capacity of the unit and the nature of the products.

Measuring and control instruments

Equipment	Use
Mechanical balance (to 100 kg)	Raw material weight, finished products in bulk
Accurate balance to 2 kg (accuracy ± 1 g)	Weighing samples, sachets, inputs (sugar, etc.)
Mercury thermometer, temperature probe	Measurement of ambient temperature or the drying air
Hair hygrometer or wall-mounted or probe hygrometer; facility for running the dryer with a psychrometer: wet and dry thermometer	Measurement of relative humidity
Refractometer	Brix measurement: level of concentration of two components of a homogeneous mixture

Measuring equipment: use of the refractometer.

Handling equipment

Equipment	Use
Tanks and vats: polythene or stainless steel	Product handling or removal of waste
Trolley	Product handling or removal of waste

Handling equipment: upstream storage rack.

Production equipment

Equipment	Use
Tables and boards for chopping (edge, drain hole, slight slope); often made locally to measure	Product preparation
Manual, foot-operated or automatic thermosealers for sachets	Packaging
Sorter for small fruit, peelers	Fruit preparation

Production equipment: (a) chopping table; (b) imported peeling machine; (c) locally made cooking vat with stirrer.

49

Chapter 4

Increasing profitability and making the right technical choices

This final section deals with aspects of a strategic nature and aims to throw light on different stages of the drying business within a small-scale unit which frequently raise questions. From the economic point of view, as regards profitability, we consider the main ways of diversifying products and improving quality. From a technical point of view, we deal with the question of technological innovation required for products based on cereals and root vegetables.

4.1 Diversifying the product range

A small-scale drying unit is often set up when a product has been identified. The product can be a raw material, such as vegetables, where the value is to be increased, or a finished product to be promoted. A business that relies on a single product is seasonal: this is frequently the case in the food-processing industry. This situation is quite acceptable if the production period is long enough, the added value from drying is high enough, and the market and market position of the business are stable enough. However, these three conditions rarely occur together. It is always possible to influence the latter two factors, but not in the short term. One solution for improving the profitability of a business can be to diversify production in order to guarantee increased productivity for technical measures introduced. The following example allows you to compare the main economic features of a single or multi-product business.

Base assumptions

The economic study is carried out on a small-scale fruit and vegetable drying business made up of small, solar dryers. The analysis covers two businesses with the same internal organization and identical equipment, but with different production.

- The first, 'Monocoq', is a single-product business which processes only one fruit, mango, for four months.
- The second, 'Multicoq', makes the most of the market gardening production period and processes potatoes and tomatoes as well as four months of mango.

The production assumptions are given in Table 4.1.

Table 4.1 Production assumptions

Component	Monocoq	Multicoq
Dried product	Mango, 4 months, 480 kg	Mango + potato, 2 months, 360 kg, and tomato, 2 months, 81 kg
Employment	Three people for 4 months (12 person-months)	Three people for 8 months (24 person-months)
Production costs	Utilities (water, electricity), packaging, transport costs proportional to quantity sold	
Capital costs	See Appendix 12	
Fixed costs	Marketing, rent	

Analyses of short- and medium-term financial results

The profitability of the business is analysed over three years. The volume of production is identical for the three years. The impact of each of these two options on the financial well-being of the drying business is analysed with the help of the two accounting methods described in Chapter 2 – the estimated operating account and the cashflow forecast. The two operating accounts are given in Appendix 12; the indicators from year 2 are shown in Table 4.2.

At this stage, taking account of the added value of the two other products, it is perfectly clear that the Multicoq option gives a better result.

Cashflow forecast

The cashflow forecast is shown in Table 4.3. At the start, the main difference between Monocoq and Multicoq relates to the amount of the

Table 4.2 Indicators (€) from operating accounts (year 2)

Component	Monocoq	Multicoq
Turnover	1555	3038
	Product diversification provides bigger income in the case of Multicoq (+95%)	
Running costs	882	1867
	The increase in running costs of 110% is not proportional to the increase in production (+91%) because of the cost of raw materials, which vary according to the product	
Gross margin	673	1172
	The addition of two extra products brings an increase of €499 even though they are less lucrative than the first	
Depreciation	196	205
	The sum allocated by Monocoq is slightly greater to take account of the frequent replacement of trays through usage	
Financing cost	238	366
	The financing costs are proportional to the level of loans (see cashflow forecast below)	
Net profit	435	601
	Product diversification provides a net profit of €166 = 38% increase	

Table 4.3 Comparison of the two cashflow forecasts (year 0)

Component	Monocoq	Multicoq
Receipts (€)		
Labour	534	534
Self-financing capacity	0	0
Loans	1981	3048
Payments		
Investments	1491	1491
Working capital	901	1864
Capital repayment	0	0
Annual cash balance	**123**	**227**

Table 4.4 Results after 3 years' operation

Component	Monocoq	Multicoq
Cashflow balance	725	1595

loan, because if the initial investment is the same, the working capital requirements depend on the level of activity. So these were counted as equivalent to the charges for the first year of production. This explains why the Multicoq loan is €1067 greater than that of Monocoq for the same amount of capital investment.

After three years (Table 4.4), in the same conditions, Monocoq makes €870 less than Multicoq, which is more that the cashflow accumulated by Monocoq activity.

This example relies on figures arising from the experiences of promoters. It shows the value of diversifying income with a view to making a return on investments. However, the best return is not always the priority, and diversification can become a means of minimizing risk. In every case, a multi-product approach is always preferable to two single product activities.

4.2 Investment in quality

A product diversification strategy, though it has the advantage of minimizing risk and increasing profitability, is not always the right solution to adopt, for one of the following reasons.

- Because the existing activity already uses the means of production for 10–12 months of the year, and this option would require over-investment.
- Because the possibilities for diversification are not sound: no value is added or the market access is difficult.

So it is preferable to improve the current activity using the tools presented in Chapter 2: economic analysis and quality management.

Use of the accounting techniques, combined with good technical knowledge of the business, enables the identification of key items that produce poorer results. The techniques for managing quality enable procedures to be improved. It is then necessary to find out whether these measures are profitable. The following example presents a comparative analysis of the workings of a business unit in two strategically opposing situations.

- 'Qualminus' represents a situation in which several items are open to improvement, but are left as they are.
- 'Qualplus' represents the same business, but with improvement in the key items from the second year.

Reference assumptions

The example represents a drying business for a product (PF) intended for export. The production season is six months of the year. The market takes up the quantities produced (demand is greater than supply). The business is equipped with eight locally produced dryers, with a unit capacity of 40 kg fresh produce per day.

The lists below summarize the assumptions taken into account (see Appendix 13).

Technical sales data

Qualminus

- Operates over 4.5 months due to 20 cycles per month.
- Produces 62 kg dried products per cycle, that is, 5580 kg/year in the third year (cruising speed).
- Sells 90 per cent of its production for export, divided into first choice (70 per cent) and second choice (30 per cent).

Qualplus

- Operates over 5.5 months due to 22 cycles per month: effort put into regularity of supply (contract with producer) which prevents breaks in production.
- Produces 64 kg per cycle, that is, 7744 kg/year (+39 per cent compared with Qualminus); checking to prevent frequent over-heating.
- Sells 95 per cent of production for export, divided between first choice (90 per cent) and second choice (10 per cent); control of process to limit second choice and losses, strict management of remaining stock (gifts, samples), sale of waste products as organic manure.

Operating costs

Qualminus

- Uses 950 kg raw materials (normal quality) to prepare 320 kg load.
- Packaging divided according to quality of finished products.
- Gas consumption equal to a return of 2.5 kWh/kg water ex-tracted.
- Employs 18 unskilled workers plus one foreman.

Qualplus

- Uses only 800 kg raw material, selection on purchase.
- Purchase price: €0.03/kg + €0.01/kg of transport: negotiation of contract with producer.
- Consumes 10 kg less gas per cycle: management of running dryer (2 kWh/kg water) plus minor technical improvement (see 'Investment', below).
- Division of staff is different: the unit favours control and can limit manpower thanks to productivity efforts.

Investment

Qualplus
The additional costs basically arise from three items.

- A larger building (+20 per cent) to work in good conditions from year 2.
- A more expensive dryer (+28 per cent) linked to technical improvements (control and regulation) which makes it possible to guarantee the quality of drying.
- Mesh flooring, with a better adapted model which meets food standards (no marking of products).

In all, this increases the investment cost by about 20 per cent, forming part of the quality cost.

Operating costs
Qualplus emphasizes control and management of the business, as follows.

- Purchasing manager to manage contracts.
- A factory manager who is better paid and more available to organize the work better.

The additional cost of €915 per year is essential for setting up a high-quality system.

Economic results

According to these assumptions, the effect of choices made on the result of the business is as follows (complete results are given in Appendix 13).

Turnover

Qualminus
With a selling price of €4.6/kg for export and €3/kg locally, annual turnover is up to €20672.

Qualplus
The weighted selling price is €0.30/kg higher. Thanks to the increase in production, turnover goes up to €32507, that is +45 per cent. To this must be added the increased value of the by-products (waste), which raises turnover by €443.

Estimated operating account
Qualplus/Qualminus comparison

- Calculating production cost shows a profit of €0.10/kg dried product for Qualplus (€1.95/kg as opposed to €2.05/kg for Qualminus).
- Added value is higher for Qualplus.
- Depreciation costs are €457 higher for Qualplus, but this represents on average only 15 per cent of added value, as opposed to 19 per cent for Qualminus.
- The net profit for Qualplus in a full output year goes up to €12195 as opposed to €6860 for Qualminus.

Working capital required (maximum in year 3) is about €1500 higher for Qualplus, taking into account the gap between the different levels of production.

Cashflow forecast
Qualplus/Qualminus comparison

- Funding: with identical capital, the total of borrowings reaches €20589 for Qualplus, as against €18293 for Qualminus. The difference finances the extra investment costs and working capital required.
- Staffing: the difference between the two, apart from the levels of investment and the working capital required, is with regard to specialized jobs. Qualplus plans two allocations (years 1 and 3) of €1520 for staff training to guarantee expertise combining quality with productivity.

- Cash balance: the first three years produce a balance, which is almost the same for Qualplus and Qualminus. However, from year 3 the efforts made by Qualplus are rewarded by a significant increase in its net balance (+ €11526 as against €4175 for Qualminus). After five years, the accumulated cash balance for Qualplus reaches €37260 or +290 per cent of that of Qualminus, thanks to the improved net trading profit.

These figures, based on realistic assumptions about the workings of small-scale business units, demonstrate the advantage for a business of embarking on a high-quality approach.

4.3 Making the right technical choices

A policy of diversification and high quality should, of course, be accompanied by suitable technical choices. This applies to both products and dryers.

New demands to be met

Until recently, the priorities as regards food-product drying in the countries of the South involved increasing the value of surplus market produce or perishable fruits, particularly for export. Today, in a number of these countries the economic advantages of drying apply not only to just the tropical fruit sector, but also to products based on cereals or root crops (flours, couscous, semolina) intended for local urban consumption. This fairly recent development requires a new approach on the part of technical advisors (NGOs, institutes) who wish to respond to the needs of businesses.

Cereals and root crops are the main products that are consumed daily, both by urban and rural people. In many countries the consumption of rice and wheat has increased greatly, to the detriment of local cereals (millet, sorghum) or root crops (cassava, yam), which are more difficult to prepare. Nevertheless, since the early 1990s the change in the parity of the CFA (African financial community) franc; the emphasis placed on increasing the value of local products; and socio-economic changes (rapid urbanization) have led to the gradual appearance of a new range of products based on grain or root crops in urban food styles: pre-cooked, dried products. *Araw* and *thiacry* in Senegal, *attiéké* in the Ivory Coast, *aklui* in Benin, sweet *dégué* in Burkina Faso, and yam in the Congo are traditional products which have been consumed in their fresh state for a very long time. In ready-to-use, dried form, which can be kept for several months, they are new products which are of increasing interest to urban households. Small-scale businesses are beginning to occupy this market niche, and drying units – mainly run by women – are gradually being set up to meet growing demand. But these production units often rapidly run into technical problems, and have difficulty finding dryers suited to their needs.

Adapting technology

Current limits and constraints
We have seen that the supply of local dryers is limited, for several reasons: demand is still weak; lack of cooperation between those with ideas and the private sector; lack of involvement of operators in the process of setting up the project; approach often limited to technical aspects, without taking account of the economic aspects.

Most drying units still use open-air drying on tiles or matting. The technical method has many disadvantages: no protection; drying uncertain in wet season; large ground surface area; low productivity. There are, of course, working dryers (see Chapter 3), but each has certain limitations.

- Solar dryers (banco dryer, data sheet A; CERER cabinet dryer, data sheet B) do not work very well (low drying speed, equipment easily damaged).
- Although air is recycled, the electric dryer prototypes produce drying costs (apart from depreciation costs) which are prohibitive for small-scale use. The electrical option for providing heating energy is risky from an economic point of view, taking into account the cost per kWh charged in most ACP countries.
- Imported, industrial dryers have good technical performance, but do not meet the needs of small-scale units. They are, nevertheless, often used in development projects supported by institutional technical cooperation, in which the promoter plays a minor role, particularly in the financing. The promoter finds himself without resources when it comes to continuing the business (lack of technical expertise, maintenance problems, etc.).
- The gas cabin dryer (ATESTA type) is largely used for drying fruits and vegetables, and sometimes cereals. Its weak points are: nominal load too low; energy cost too high; drying method (natural convection, beating) too limited.

The difficulties in adapting the current supply of technical equipment for drying cereal products are explained by the techno-economic demands of this sector.

- The level of small-scale production is very varied, but is increasing. A production aim of 100 kg dried products per day cannot be met by small dryers or industrial dryers.
- Operators, who are not technical experts, have requirements to which it is not always easy to respond: the dryer has to be working all year, with or without sun; has to have a reasonable investment cost; and must be technically manageable.
- The total cost of drying should be limited to 80 CFA francs, or 50 CFA francs/kg dried product, in order to maintain a sales price which is attractive to mass consumption: this ceiling leaves no room for dryers whose energy management has not been thought through from the outset (simple dryers are often energy 'guzzlers'), or imported dryers whose financing costs are prohibitive.

Recent developments
For several years, GERES has contributed to the development of new models of dryers, particularly for use in small-scale drying units. It does not have a 'miracle cure' to respond to these needs, but it does adopt an approach which is both pragmatic and empirical.

- Pragmatic – it is in direct contact with the economic operators who have to invest in the development phase while taking a share of the risks: there are very few who accept this gamble.
- Empirical – it does not try to create models, but rather to develop what is already there in order to innovate.

Innovating does not involve inventing a new dryer, but instead adapting an existing one (whether an imported or local model) in accordance with certain specifications. These specifications impose technical choices: use of mechanical air ventilation in order to reach desired production levels and control flow of air in relation to drying speed; provision of rational energy supply to carry out drying in all seasons, while avoiding excessive consumption. Apart from the technical aspects, innovating also involves ensuring from the outset that economic criteria have been taken into account and that conditions are in place for local control of expertise and manufacturing.

The experimentation currently in progress is directed towards two lines of research.

- An air-recycling tunnel dryer model (Cartier type) for units with production of above 200 kg dried products per day.
- A mixed, solar/gas dryer model for units with 50–200 kg dried products per day.

Appendix 1

Example of a semi-directive questionnaire for an 'organic' product consumer

Details of respondent (sex, age, address, socio-professional group):

_____

_____

Date and place of purchase:

_____

Brand purchased:

_____

Quantities:

_____

Frequency, reason for purchase of this product:

_____

_____

Criteria for the choice of product:

❑ Price   ❑ Ease of use

❑ Taste quality   ❑ Brand

Expectations and needs satisfied by consumption:

❑ Good for health   ❑ Nutrition

❑ Natural

Defects of product purchased:

❑ Price   ❑ Packaging

❑ Use

Expressed desire for improvements:

_____

_____

Knowledge of competitor products:

_____

_____

Use of substitute products:

_____

_____

Appendix 2

Example of questionnaire to target relative non-consumers

Date: _____

Survey reference: _____

Profile of the customer (sex, age, address, socio-professional group):

_____

Do you know of the product, 'Product X'?

_____

Do you purchase it?

❑ Yes: see questionnaire in Appendix 1

❑ No: continue

Do you purchase products Y^*
[*products with similar characteristics]

_____

What faults do you find with Product X?

_____

_____

What differences do you see between products X and Y?

_____

_____

If the faults of X were corrected, would you buy it?

_____

_____

Where would you wish to purchase product X?

_____

_____

Appendix 3

Doing your own export analysis

Question	Strengths	Weaknesses
At production level		
Are my production capacities able to respond to increased demand?		
Is my equipment working?		
At technical level		
Do I have a considerable expertise in drying?		
Am I in a position to propose a whole range of dried products?		
Do I have to reorganize my range of products and production processes in accordance with demand from target markets?		
Do I have a quality control system within the business?		
At commercial level		
Do I have a good reputation on the current local market?		
Do I have any knowledge of the markets targeted, level of competition, distribution networks or should I carry out a market survey?		
What partner will I have?		
Do I know enough about current regulations to export to the countries concerned?		
At financial level		
Do I have the means to break new ground?		
If I have to reorganize my production, will it cost me too much?		
Do I have the capacity to absorb the financial costs associated with export operations?		
Am I in a position to evaluate the development of turnover and profit margins?		
How will I finance my investments?		
At human resources level		
Are my staff qualified enough?		
Will I be able to pay for training courses?		

Appendix 4

The main 'incoterms' (international commercial terms) as defined by the International Chamber of Commerce

Incoterm	Seller	Buyer
Ex-Works (on leaving factory)		Must take delivery as soon as it is available
FAS (free alongside ship) + name of a port	Not responsible for export formalities but provides the necessary documentation	Responsible for the export formalities; chooses the ship
FOB (free on board) + name of a port	Exports the merchandise, delivers it on board the ship and provides the documents certifying its delivery	Responsible for the choice of ship, bears the expense of obtaining and the cost of the bill of lading (shipping documents)
CFR (cost and freight) + name of a port	Exports the merchandise, delivers it to the ship, provides the bill of lading, is no longer responsible for the merchandise once it is on board, but pays the freight and unloading at destination if included in the freight; also called 'CFR landed'	Responsible for the merchandise once it is on board; responsible for importing
CIF (cost, insurance and freight) + port	Same as CFR + the seller must provide insurance to cover the CIF price + 10%, up to delivery	Same as CFR + the buyer pays insurance over and above the so-called 'FPA' risks (such as the pilferage, leakage, packing, condensation in the hold, etc.)

Appendix 5

Instruments of payment

Cheque
Advantage: simplicity.
Disadvantage: payment left to goodwill of customer.

Transfer
The person owing money instructs their bank to pay the person to whom they owe money. There is no written text as for the cheque. It can be very quick, if it is SWIFT (interbank network) or by telex.
Advantage: simple, quick, safe.
Disadvantage: payment left to goodwill of customer.

Bill of exchange
The drawer gives a written order to those being paid to pay on demand or when due. This is an instrument of payment and credit.
Advantage: simple, possible discount.
Disadvantage: risk of non-payment and non-transfer if bill of exchange is not endorsed by the bank of those being paid.

Promissory note
The person making payment promises in writing to pay the person owed an amount fixed to a date and precise place.
Advantage: simple, possible discount.
Disadvantage: non-payment, non-transfer, writing errors.
The organization system put in place to collect the instrument of payment allows one to obtain certain guarantees of payment in exchange for a quality product.

Cash on delivery
Seller instructs the freight-forwarding agent to deliver the goods on receipt of payment.
Advantage: safe for seller.
Disadvantage: reluctance on the part of forwarding agents to use this process.

Delivery on presentation of document
Seller instructs bank to ensure that export documents reach purchaser upon payment or acceptance of bill.
Advantage: low cost.
Disadvantage: non-payment on expiry of bill.

There are other, more sophisticated instruments of payment, such as different credit documents (revocable, irrevocable, confirmed), which are little used in the fresh or dried fruit trade.

Appendix 6

Example of specification for obtaining a quotation for procuring a dryer

FAX

From: (*Business name*)   To: (*Equipment provider*)
          Ms/Mr (*entrepreneur*)   Ms/Mr (*seller*)

Dear *Sir/Madam*,

SUBJECT: REQUEST FOR PRICE INFORMATION

We are a food-drying business based at (*location of business*). I am looking for a food dryer to meet our needs as follows:

- To process: red chilli peppers, *pili pili* type whole, length 2–3 cm, diameter 1 cm, initial moisture content about 75 per cent.
- Market: European export market. Colour retention essential. Final moisture content: 9 per cent.
- Weekly production: 1 tonne fresh produce. Medium-term forecast: 2 tonnes fresh produce.
- Electrical supply: 24 hours per day. Gas: available in bottles of 9–36 kg.
- Ambient limiting conditions: 15°C, 85 per cent relative humidity.

I should be grateful if you would send me your best offer with the corresponding technical specification.
Yours faithfully,

(*Name*)

(*Business*)

Appendix 7

Introducing the quality module into the training session 'Setting up Food-drying Businesses in Africa', Ouagadougou, December 1998

Concept of food safety (FAO, 1997)

In the developing world, millions of infants die because of food and water contamination, often due to poor hygiene conditions in the home.

Every year, poor handling after harvest and contamination lead to the loss of millions of tonnes of food.

On a worldwide scale, losses of food due to contamination from fungal toxins is up to about 100 million tonnes per year.

Exports sent back because of their poor quality cost countries millions of dollars.

That is why it is essential to guarantee food safety and quality through world trade.

Furthermore, controlling food quality and safety is important for good development of the food industry of a country, because it enables:

- jobs to be created
- incomes to be increased
- people's nutrition to be improved.

To achieve fairness between nations and protect consumers, there must be consensus about what is allowed and what is forbidden.

Food quality and safety standards should be adapted to developing lifestyles, particularly urbanization, farming practices, food processing and marketing.

The perception of quality in developing countries

Obstacles to improving quality in the industry in developing countries include a number of misunderstandings, as follows.

First misunderstanding: 'better quality costs more'
However, new findings about developing quality (in research and development) and processes have shown that good quality does not always cost more.

Second misunderstanding: 'emphasizing quality brings lower productivity'
Previously this was true because strict control criteria led to rejection of much of the production. Now, emphasis is on prevention during design and manufacturing so that faulty goods are not produced.

Third misunderstanding: 'the workforce is totally responsible for poor quality'
Manufacturers in developing countries often attribute the poor quality of their products to the fact that their workers are not sufficiently aware of quality, and do not consider their work as of value.

Rather than making accusations, manufacturers should investigate the failings in their management systems.

Observations on the uneven nature of quality in finished products from developing countries

Many private companies in developing countries belong to families. Consequently, the authority for taking decisions is centralized within the family, and information does not necessarily reach less senior members. Hence there may be significant departures from the norm, and a lack of regular quality in finished products.

Businesses in developing countries face difficulties such as limited finances for energy, transport and communications.

Management may be concerned with producing in quantity, even if they risk not keeping to production standards: this can have serious consequences in the long term for export markets.

It is important that developing countries should look further at this management culture centred on quantity and focused on the technical expertise of certain key members of staff.

Appendix 8

Typical layout of a small-scale drying unit

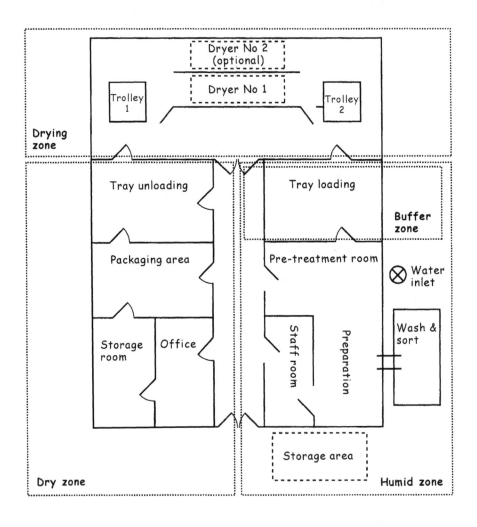

Appendix 9

Forecast business performance

Heading	Years (0,1,2...N)
Turnover (A)	Manufacturing revenue (value of products and by-products)
Operating costs (B)	Costs directly related to quantity produced (raw materials, energy, inputs, consumables, labour)
Gross margin, GM = A – B	Intermediate result relating to production cost
Fixed costs (C)	Operating costs regardless of quantity (location, rentals, management)
Added value, VA = GM – C	Significant intermediate profit from processing income
Taxes (except on profit) (D)	Deductions levied on businesses by tax authorities
Trading profit, TP° = VA – D	Economic interest value of business in local context
Depreciation charges (E)	Depreciation is an important concept with regard to gradual reduction in capital assets (buildings, equipment) due to wear and tear, obsolescence; this allowance is a provision for amount of depreciation
Gross profit, GP = TP – E	Economic value taking into account material means used
Interest charges (F)	Expenses relating to interest on loans taken out for production
Pre-tax profit, PTP = GP – F	Economic value of business within the business unit involved
Percentage of tax on profit (G)	Contribution of the business to workings of the State, proportional to profits
Net profit, NP = PTP – G	Net profit from production business taking into account financial means and environment
Cashflow, CAF = NP + E	Funds readily available to the business for the coming financial year, ensuring continuing operation

Appendix 10

Cashflow forecast, receipts – payments

Heading	Years (0,1,2...N)
Total receipts (A)	Year 0, start of investments Year 1, start of production
Personal contribution	Capital invested by promoter and partners
Sales	Yearly earnings
Term loans	Loans from bank credits made to complete the resources
Total payments (B) Investment	Total sum invested in the enterprise
Working capital requirements	Sum of money the enterprise requires for production, taking into account delays in clients settling bills; working capital is calculated as a percentage of the costs to be covered in the current year, and is covered each year by the cashflow – the level takes account of the development of requirements
Loans	If the financing costs are part of the estimated operating account, the capital repayment appears in this table for each year that the loan is outstanding (whether or not deferred); the bank claims it according to a repayment schedule (constant annuity)
Dividends	Each shareholder in the business receives a share of the profits every year
Net cash balance (A – B)	Cash sum available following the annual accounts – in year 0, the balance must be positive in order to start trading; subsequently, if the balance is negative, the business must dig into its previously accumulated reserves
Cumulative cash balance	Final indicator of the viability of the business – it must never be negative or the business will cease (bankruptcy), unless they increase receipts (inject new money), or reschedule (arranged with the bank)

Appendix 11

Supply chains of dried fruit in Europe

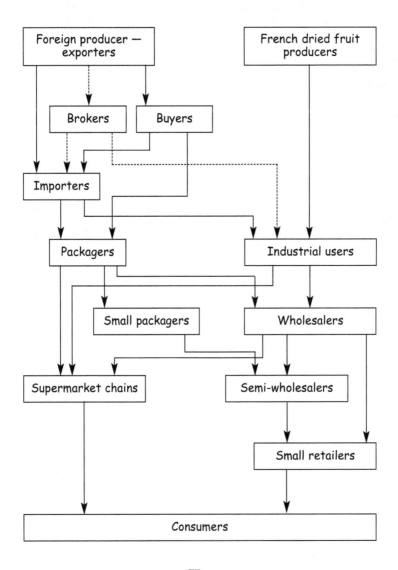

Appendix 12

Economic analysis for Monocoq and Multicoq

Investment	Quantity	Unit value (€)	Total cost (€)	First year of investment	Duration (years)	Depreciation period (years)
Building the unit			950			
Construction in resistant materials	25	38/m²	950	0	11	10
Drying equipment			453			
Dryer No. 3	6	73/m²	439	0	7	6
Paint	1	5/m²	5	2	2	1
Tray	1	9/m²	9	2	2	1
Associated equipment			85			
Knives	7	2.29	16	0	3	2
Tables	1	15	15	0	5	4
Welder	1	38	38	0	6	5
Bucket	7	2.29	16	0	3	2
Total investment			1488			

Estimated operating account and cashflow forecast for Monocoq and Multicoq

	Start	Year 1	2	3
(I) Monocoq				
(a) Estimated operating account				
Sales		1555	1555	1555
Potatoes sold (kg)		0	0	0
Mangoes sold (kg)		480	480	480
Tomatoes sold (kg)		0	0	0
Expenses	0	864.7	864.5	864.5
Inputs		440	440	440
Utilities		51	51	51
Transport		7.3	7.3	7.3
Labour		366	366	366
Miscellaneous		0.4	0.2	0.2
Gross margin		690.3	690.5	690.5
Depreciation		196	196	194
Financing costs		238	238	214
Profit before tax		256.3	256.5	282.5
Net profit		256.3	256.5	282.5
Max. self-financing (cashflow)		494.3	494.5	496.5
Cumulative profit		**256.3**	**512.8**	**795.3**
(b) Cashflow forecast				
Receipts	2516	494.3	494.5	496.5
Initial support	534	0	0	0
Max. self-financing	0	494.3	494.5	496.5
Loan	1982			
Payments	2389	196	276	278
Investments				
Construction	950			
Dryer No. 3	453	0	14	0
Minor items	85	0	0	32
Working capital	901	0	0	0
Capital repaid	0	196	262	246
Annual cash balance	**127**	**298.3**	**218.5**	**218.5**
Cumulative balance	**127**	**425.3**	**643.8**	**862.3**

Continued overleaf

Continued from previous page

(II) Multicoq
(a) Estimated operating account

Sales	0	3038	3038	3038
Potatoes sold (kg)		360.8	360.8	360.8
Mangoes sold (kg)		480	480	480
Tomatoes sold (kg)		81	81	81
Expenses	0	1886	1867	1867
Inputs		1018	1018	1018
Utilities		84	84	84
Transport		14	14	14
Labour		732	732	732
Miscellaneous		38	19	19
Gross margin	0	1152	1171	1171
Depreciation	0	205	205	203
Financing costs	0	366	366	330
Profit before tax	0	581	600	638
Net profit	0	581	600	638
Max. self-financing (cashflow)	0	779	797	833
Cumulative profit	**0**	**779**	**1576**	**2409**

(b) Cashflow forecast

Receipts	3582			
Initial support	534	1031	1050	1086
Max. self-financing	0	779	797	833
Loan	3049			
Payments	3371	302	352	411
Investments				
Construction	953			
Dryer No 3	453	0	14	0
Minor items	85	0	0	32
Working capital	1880	0	0	0
Capital repaid	0	302	338	379
Annual cash balance	**211**	**477**	**445**	**422**
Cumulative balance	**211**	**688**	**1143**	**1565**

Appendix 13

Economic analysis of Qualminus and Qualplus

Techno-economic data sheet for product 'PF Minus'

Nominal data		*Target client*
Nominal target (kg)	5580 kg	Grade 1 Export 70%
Nominal year	3	Grade 2 Local 30%
No. months of production in year 3 to achieve 5580kg	4.5	

Unit production cycle		**List of operations in a cycle and saleable by-products**		
Cycle time:	22 hours	***Process***	***By Product***	
		Name	***Name***	***%/PF Minus***
		Drying	–	–
Quantity of PF Minus per cycle	62			
Number of cycles per year in nominal year 3	90			

Five-year production table for PF Minus

				Year		
	0	*1*	*2*	*3*	*4*	*5*
Scale of production (%)	0	60	80	100	100	100
Quantity produced by						
PF Minus (kg)	0	3348	4464	5580	5580	5580
Level of sales (%)	90		Level of unsold: 10			
Total sales (kg)						
Export	0	2109.2	2812.3	3515.4	3515.4	3515.4
Local	0	904.0	1205.3	1506.6	1506.6	1506.6
Saleable by-products	0	0	0	0	0	0

Operating costs of production in a cycle

PF Minus	Quantity consumed per cycle	Unit cost €
Raw material (kg)	950	0.04/kg
Inputs		
Export package	182	0.12
Local package	81	0.06
Box	6	0.69
Bleach (litres)	0.25	6.10/litre
Utilities		
Gas (kg)	51	0.46/kg
Water (m³)	0.5	0.76/m³
Labour (person days)		
Manual workers	18	1.68/person day
Supervisor	2	3.35/person day
Transport		
Other		

Techno-economic data sheet for product 'PF Plus'

Nominal data		Target client
Nominal target	7744 kg	Grade 1: Export 90%
Nominal year	3	Grade 2: Local 10%
Number of months of production in year 3 to achieve 7744kg	5.5	

Unit production cycle

List of operations in a cycle and saleable by-products

		Process Name	By Product Name	%/PF Plus
Cycle time:	22 hours	Trimming	Compost	75%
Quantity of PF Plus per cycle	64			
Number of cycles per year in nominal year 3	121			

Five-year production table for 'PF Plus'

	Year 0	1	2	3	4	5
Scale of production (%)	0	50	80	100	100	100
Quantity produced by PF Plus (kg)	0	3872	6195	7744	7744	7744
Level of sales (%)	95		Level of unsold: 5			
Total sales (kg)						
Export	0	3310	5297	6621	6621	6621
Local	0	368	589	736	736	736
Saleable by-products	0	2904	4646	5808	5808	5808

Operating costs of production in a cycle

PF Plus	Quantity consumed per cycle	Unit cost (€)
Raw material (kg)	800	0.03/kg
Inputs		
Export package	222	0.12
Local package	31	0.06
Box	7	0.69
Bleach (litres)	0.25	6.10/litre
Utilities		
Gas (kg)	41	0.46/kg
Water (m³)	0.5	0.76/m³
Labour (person days)		
Manual workers	16	1.83/person day
Supervisor	3	3.66/person day
Transport (kg)	800	0.007/kg
Other		

Qualminus list of capital investments

	Quantity	Unit value (€)	Total cost (€)	Lifetime	Depreciation period	Investment year
Structure						
Building (m²)	150	61/m²	9150	15	14	0
Total			**9150**			
Processing Equipment						
Dryer	8	534	4272	8	7	0
Trays	16	6.10	98	2	1	1
Mesh flooring (m²)	60	7.62/m²	457	2	1	0
Total			**4827**			
Associated Equipment						
Furniture	5	152	760	6	5	0
Gas welding	1	457	457	4	3	0
Total			**1217**			
Minor Items						
Miscellaneous	5	46	230	3	2	0
Total			**230**			
Instrumentation						
Thermometer	1	46	46	4	3	0
Total			**46**			
Other			0			
Total			**15470**			

Qualplus list of capital investments

	Quantity	Unit value (€)	Total cost (€)	Lifetime	Depreciation period	Investment year
Structure						
Building (m²)	150	61/m²	9150	15	14	0
Extension (m²)	30	61/m²	1830	13	12	2
Total			**10 980**			
Processing equipment						
Dryer	8	686	5488	8	7	0
Trays	16	6.10	98	2	1	1
Mesh flooring (m²)	60	12.20/m²	732	2	1	0
Total			**6318**			
Associated equipment						
Furniture	5	152	760	6	5	0
Gas welding	1	457	457	4	3	0
Total			**1217**			
Minor Items						
Miscellaneous	5	46	230	3	2	0
Total			**230**			
Instrumentation						
Thermometer	1	46	46	4	3	0
Total			**46**			
Other			0			
Total			**18 791**			

Qualminus: turnover of all finished products in a nominal year

	Unit price (€/kg)	Annual quantity (kg)	Annual total (€)
Finished Product: PF Minus			
Grade 1	4.57	3515.4	16 066
Grade 2	3.057	1506.6	4606
Saleable by-products	0	0	0
Total			**20 672**

Qualminus: details of operating costs for the finished product PF Minus

	Annual quantity	Unit cost (€)	Total (€)	Percentage
Raw material (kg)	85 500	0.038/kg	3249	27.5
Total			**3249**	**27.5**
Inputs				
Export package	16 380	0.12	1966	16.6
Local package	7290	0.06	437	3.7
Box	540	0.69	373	3.2
Bleach (litres)	23	6.10/litre	140	1.2
Total			**2916**	**24.7**
Utilities				
Gas (kg)	4961	0.46/kg	2282	19.3
Water (m^3)	45	0.76/m^3	34	0.3
Total			**2316**	**19.6**
Labour (person days)				
Manual workers	1620	1.68/ person day	2722	23.1
Supervisor	180	3.35/ person day	603	5.1
Total			**3325**	**28.2**
Transport			0	0
Total				
Other				
Total				
Final total			**11 806**	**100**

Qualminus: overhead costs (€)

	Quantity	Unit value	Total	Percentage
Rent			0	0
Transport				
Total			**0**	**0**
Administrative staff				
Director	6	107/person month	640	100
(person months)				
Total			**640**	**100**
Marketing costs			0	0
Total			**0**	**0**
Miscellaneous			0	0
Total at full output			**640**	**100**

Qualminus: level of overhead costs over five years (%)

	Year 0	1	2	3	4	5
Level	0	100	100	100	100	100

Qualplus: turnover of all finished products in a nominal year

	Unit price (€)	Annual quantity (kg)	Annual total (€)
Finished product: PF Plus			
Grade 1	4.57/kg	6621.1	30 258
Grade 2	3.057/kg	735.7	2249
Saleable by-products			
Compost	0.0762/kg	5808	443
Total			**32 950**

84

Qualplus: details of operating costs for the finished product PF Plus

	Annual quantity	Unit cost (€)	Total (€)	Percentage
Raw material (kg)	96 800	0.0381/kg	3688	26.0
Total			**3688**	**26.0**
Inputs				
Export package	21 190	0.12	2543	17.9
Local package	3751	0.06	225	1.6
Box	847	0.69	584	4.1
Bleach (litres)	23	6.10/litre	140	1.0
Total			**3492**	**24.6**
Utilities				
Gas (kg)	4590	0.46/kg	2111	14.9
Water (m³)	45	0.76/m³	34	0.2
Total			**2145**	**15.1**
Labour (person days)				
Manual workers	1936	1.83/person day	3543	25.0
Supervisor	363	3.66/person day	1329	9.4
Total			**4872**	**34.3**
Transport			0	0
Total			**0**	
Other			0	
Total			**0**	
Final total			**14 197**	**100**

Qualplus: overhead costs (€)

	Quantity	Unit value	Total €	Percentage
Rent			0	0
Transport				
Total			**0**	**0**
Administrative staff (person months)				
Director	9	122/ person month	1098	71
Assistant	6	76.22/ person month	457	29
Total			**1555**	**100**
Marketing costs			0	0
Total			**0**	**0**
Miscellaneous			0	0
Total at full output			**1555**	**100**

Qualplus: level of overhead costs over five years (%)

	Year 0	1	2	3	4	5
Level	0	100	100	100	100	100

Qualminus: estimated accounts (€)

	Year 0	1	2	3	4	5
Products						
Finished products sold	0	12400	16536	20670	20670	20670
By-products sold	0	0	0	0	0	0
Total	0	12400	16536	20670	20670	20670
Operating costs						
Raw materials	0	1955	2607	3258	3258	3258
Inputs	0	1770	2360	2950	2950	2950
Utilities	0	1280	1707	2133	2133	2133
Labour	0	1992	2656	3320	3320	3320
Transport	0	0	0	0	0	0
Other	0	0	0	0	0	0
Total		6997	9330	11661	11661	11661
Gross margin		5403	7206	9009	9009	9009
Overhead costs						
Rent	0	0	0	0	0	0
Transport	0	0	0	0	0	0
Admin/personnel	0	640	640	640	640	640
Marketing costs	0	0	0	0	0	0
General	0	0	0	0	0	0
Miscellaneous	0	0	0	0	0	0
Total	0	640	640	640	640	640
Added value	0	4763	6866	8369	8369	8369
Tax due	0	0	0	0	0	0
Tax as percentage	0	0	0	0	0	0
Gross profit	0	4763	6866	8369	8369	8369
Depreciation allowance	0	1502	1795	2041	1627	2155
Operating margin	0	3261	5071	6328	6742	6214
Interest on loans	0	2424	2330	2074	1779	1440
Profit before tax	0	837	2741	4254	4963	4774
Tax on profits	0	0	0	0	0	0
Net profit	0	837	2741	4254	4963	4774
Cashflow	0	2341	4236	6294	6589	6928
Working capital requirements						
Level of working capital	0%	100%	100%	90%	80%	80%
Annual working capital requirement	0	7637	9970	11071	9841	9841

Qualplus: estimated accounts (€)

	Year 0	1	2	3	4	5
Products						
Finished products sold	0	16 261	26 018	32 522	32 522	32 522
By-products sold	0	221	354	443	443	443
Total	**0**	**16 482**	**26 372**	**32 965**	**32 965**	**32 965**
Operating costs						
Raw materials	0	1476	2361	2951	2951	2951
Inputs	0	2135	3416	4270	4270	4270
Utilities	0	1157	1852	2315	2315	2315
Labour	0	2435	3896	4870	4870	4870
Transport	0	369	590	738	738	738
Other	0	0	0	0	0	0
Total		**7572**	**12 115**	**15 144**	**15 144**	**15 144**
Gross margin		**8910**	**14 257**	**17 821**	**17 821**	**17 821**
Overhead costs						
Rent	0	0	0	0	0	0
Transport	0	0	0	0	0	0
Admin/personnel	0	1555	1555	1555	1555	1555
Marketing costs	0	0	0	0	0	0
General	0	0	0	0	0	0
Miscellaneous	0	0	0	0	0	0
Total	**0**	**1555**	**1555**	**1555**	**1555**	**1555**
Added value	**0**	**7355**	**12 702**	**16 266**	**16 266**	**16 266**
Tax due	**0**	**0**	**0**	**0**	**0**	**0**
Tax as percentage	**0**	**0**	**0**	**0**	**0**	**0**
Gross profit	**0**	**7355**	**12 702**	**16 266**	**16 266**	**16 266**
Depreciation allowance	**0**	**1600**	**2350**	**1925**	**2320**	**2405**
Operating margin	**0**	**5755**	**10 352**	**14 341**	**13 946**	**13 861**
Interest on loans	**0**	**2736**	**2632**	**2339**	**2002**	**1614**
Profit before tax	**0**	**3019**	**7720**	**12 002**	**11 944**	**12 247**
Tax on profits	**0**	**0**	**0**	**0**	**0**	**0**
Net profit	**0**	**3019**	**7720**	**12 002**	**11 944**	**12 247**
Cashflow	**0**	**4620**	**10 070**	**13 928**	**14 265**	**14 653**
Working capital requirements						
Level of working capital	**0%**	**100%**	**85%**	**75%**	**75%**	**75%**
Annual working capital requirement	**0**	**9127**	**11 619**	**12 524**	**12 524**	**12 524**

Qualminus: cashflow forecast

	Year 0	1	2	3	4	5
Receipts						
Personal equity	6098	0	0	0	0	0
Cashflow available	0	2341	4236	6294	6589	6928
Loan 1	13720	0	0	0	0	0
Loan 2	4573	0	0	0	0	0
Aid and subsidies	0	0	0	0	0	0
Total	**24391**	**2341**	**4236**	**6294**	**6589**	**6928**
Payments						
Investments	15366	98	457	326	960	98
Scheduling of working capital	7637	2332	1102	0	0	0
Repayment of loan 1	0	782	876	981	1098	1230
Repayment of loan 2	0	0	890	1041	1218	1425
Exceptional expenses	0	0	0	0	0	0
Dividends	0	0	0	0	0	0
Total	**23003**	**3212**	**3325**	**2348**	**3276**	**2753**
Net cashflow balance	**1388**	**−871**	**911**	**3946**	**3313**	**4175**
Cumulative cash balance	**1388**	**517**	**1428**	**5374**	**8687**	**12862**

Qualplus: cashflow forecast

	Year 0	1	2	3	4	5
Receipts						
Personal equity	6098	0	0	0	0	0
Cashflow available	0	4620	10070	13928	14265	14653
Loan 1	15254	0	0	0	0	0
Loan 2	5335	0	0	0	0	0
Aid and subsidies	0	0	0	0	0	0
Total	**26687**	**4620**	**10070**	**13928**	**14265**	**14653**
Payments						
Investments	16905	98	1829	1058	549	98
Scheduling of working capital	9127	2493	905	0	0	0
Repayment of loan 1	0	869	973	1089	1220	1367
Repayment of loan 2	0	0	1038	1214	1421	1662
Exceptional expenses	0	1524	0	1524	0	0
Dividends	0	0	0	0	0	0
Total	**26032**	**4984**	**4745**	**4885**	**3190**	**3127**
Net cashflow balance	**655**	**−364**	**5325**	**9043**	**11075**	**11526**
Cumulative cash balance		**291**	**5616**	**14659**	**25734**	**37260**

Appendix 14

Useful contacts

Africa

Benin

Centre Songhaï, M. Ahouansou (Dépt méchanique), PO Box 597, Porto Novo
Fax: (219) 22 50 50, e-mail: songhai.benin@intnet.bj
Support Consultants, Technical Centre
Iceberg Sarl, M. Mekpato, 03 PO Box 4300, Cotonou
Fax: (229) 33 35 13
Support Consultants

Botswana

Botswana National Food Technology Research Centre (NFTRC), PO Box 008, Kanye, Botswana
Fax: (267) 34 07 13
Research Centre

Burkina Faso

ABAC-CAA, M. Traoré, 01 PO Box 4071, Ouagadougou 01
Fax: (226) 36 02 18, e-mail: abac@fasonet.bf
Support Consultants
ATESTA, M. Guissou, 01 PO Box 3306, Ouagadougou 01
Fax: (226) 30 23 99
Support Consultants
CEFOC, M. Galland, 01 PO Box 594, Ouagadougou 01
Fax: (226) 31 92 26
Training Centre
EIER, M. Coulibaly, 03 PO Box 7023, Ouagadougou 03
Fax: (226) 31 27 24
Training School
IRSAT, 03 PO Box 7047, Ouagadougou 03
Research Centre
SICAREX, M. Badini, 01 PO Box 2625, Ouagadougou 01
Fax: (226) 31 28 53
Support Consultants

Burundi

CNTA, M. Gikota, PO Box 557, Bujumbura
Fax: (257) 22 24 45
Technical Centre

Cameroon

AGRO-PME, M. Monkam, PO Box 10087, Yaoundé
Fax: (237) 23 96 92, e-mail: agro_pme.yde@comnet.cm
Support Consultants
APICA, M. Eberard, PO Box 403, Douala
Fax: (237) 37 04 02
Technical Centre

Congo

AGRICONGO, M. Bouka, PO Box 14574, Brazzaville
Fax: (242) 94 41 45, e-mail: agricongo_pnr@compuserve.com
Support Consultants, Technical Centre
FJEC, M. Ntitie, PO Box 2080, Brazzaville
Support Consultants

Ethiopia

Food Research and Development Centre, Ethiopian Food Corporation,
PO Box 5688, Addis Ababa
Research Centre

Ivory Coast

I2T, M. Guefala, 04 PO Box 1137, Abidjan 04
Fax: (225) 27 90 51
Technical Centre

Ghana

Technology Consultancy Centre (TCC), University of Science and
Technology, Kumasi
Technical Centre, Research Centre

Guinea

APEK, M. Saco, PO Box 71 Kindia
Fax: (224) 61 09 60
Support Consultants

Kenya

Department of Food Science and Post Harvest Technology, Jomo Kenyatta University, PO Box 62000, Nairobi
Tel: (254) 1 51 52181
Fax: (254) 1 51 52164/52255, e-mail: fsptiku@arcc.or.ke
Training Centre

Madagascar

CITE, M. Beville, PO Box 74, Antananarivo 101
Fax: (261) 20 22 226 69
Information Centre

Nigeria

Department of Food Science & Technology, Federal University of Technology, PMB 1526, Owerri
Training Centre
International Institute of Tropical Agriculture (IITA), PMB 5320, Ibadan
Research Centre

Senegal

ENDA-GRAF, Khanata Sokona, PO Box 13069, Dakar
Fax: (221) 827 32 15, e-mail: graf@enda.sn
Support Consultants
Gret Sénégal, Cécile Broutin, PO Box 10422, Dakar
Fax: (221) 821 98 14, e-mail: gretsn@arc.sn
Support Consultants

Uganda

Department of Food Science & Technology, Makerere University, PO Box 7062, Kampala
Training Centre

Zambia

Small Scale Industries Association, PO Box 37156, Lusaka
Tel: (260) 1 288434/252150
Information Centre

Zimbabwe

ITDG Zimbabwe, PO Box 1744, Harare
Tel: (263) 4 91 403896
Fax: (263) 4 669773, e-mail: itdg@internet.co.zw
Support Consultants

Caribbean and Pacific

Antigua

Chemistry & Food Technology Division, Ministry of Agriculture, Fisheries and Lands, Dunbars, Friars Hill Road, St John's
Tel: (1 268) 462 4502
Fax: (1 268) 462 6281, e-mail: moa@candw.ag
Training Centre

Barbados

Ministry of Agriculture and Rural Development, Graeme Hall, Christ Church
Tel: (1 246) 428 4150/0061
Fax (1 246) 428 0152
Training Centre

Dominica

Produce Chemist's Laboratory, Ministry of Agriculture and the Environment, Botanical Gardens, Roseau
Tel: (1 767) 448 2401 ext. 3426
Fax: (1 767) 448 7999, e-mail: parbel@hotmail.com
Training Centre

Grenada

Produce Chemist's Laboratory, Tanteen, St George's
Tel: (473) 440 3273/0105
Fax: (473) 440 3273, e-mail: guimacel@caribsurf.com
Training Centre

Guyana

Agricultural Projects Unit, Ministry of Agriculture, Regent Street/Vissengen Road, Georgetown
Tel (592) 2 60393
Fax: (592) 2 75357, e-mail: nisa@sdnp.org.gy
Training Centre

Jamaica

Food Technology Institute, Scientific Research Council, Hope Gardens, PO Box 350, Kingston 6
Tel: (1 876) 977 9316
Fax: (1 876) 977 2194, e-mail: fithead@cwjamaica.com
Research Centre

Applied Food and Chemistry Department of Chemistry, University of the West Indies, Mona, Kingston 7
Tel: (1 876) 927 1910,
Fax: (1 876) 977 1835, e-mail: dminott@uwimona.edu.jm
Research Centre, Training Centre

Papua New Guinea

Appropriate Technology Development Division, Ministry of Agriculture, Fisheries and Land, PO Box 793, Lae
Research Centre

Saint Kitts

St Kitts–Nevis Multipurpose Laboratory, Department of Agriculture, PO Box 39, Basterre
Tel: (1 869) 465 5279
Fax: (1 869) 465 3852, e-mail: mplbos@caribsurf.com
Training Centre

Saint Lucia

Produce Chemist's Laboratory, Research and Development Division, Ministry of Agriculture, Fisheries and Forestry, Block A, Waterfront, Castries
Tel: (1 758) 450 2375
Fax: (1 758) 450 1185, e-mail: research@slumaffe.org
Training Centre

Surinam

Agricultural Experimental Station, Ministry of Agriculture, Animal Husbandry and Fisheries, Hetitia Vriesdelaan #10, Paramaribo
Tel: (597) 472442
Fax: (597) 470301, e-mail: seedunit@sr.net
Training Centre

Trinidad and Tobago

Department of Food Production, Faculty of Agriculture and Natural Sciences, University of the West Indies (UWI), St Augustine
Tel: (1 868) 662 2002/645 3232/4 ext. 2110/2090
Fax (1 868) 663 9686, e-mail: istre@carib-link.net
Training Centre

Europe

Germany

MAURER, Obere Rheinstrasse 41, 78479 Reicherau
Fax: (49) 75 34 808
Drying Equipment
Hohenheim University, M. Mühlbauer, Garßenstrasse 9, D-70599 Stuttgart
Fax: (49) 711 459 32 98, e-mail: muelbauer@ats.uni-hohenheim.de
Research
Innotech, Hohenheim University, Institute for Agricultural Engineering in
the Tropics, Brandenburger Strasse 2, D-71229, Leonberg
Fax: (49) 07031 74 47 41, e-mail: Innotech.ing@t-online.de
Research Centre

France

CIRAD SAR, M. Meot, PO Box 5035, 34032 Montpellier Cedex 1
Fax: (33) 4 67 61 12 23, e-mail: meot@cirad.fr
Research
GERES, 2 cours Maréchal Foch, 13400 Aubagne
Tel: (33) 442 18 55 88
Fax: (33) 442 03 01 56, e-mail: geres@worldnet.fr
Gret, Martine François, 213, rue Lafayette, 75010 Paris
Fax: (33) 1 40 05 61 61, e-mail: françois@gret.org
Support Consultants
SOLAGRO, M. Bochu, 229 Avenue Muret, 31300 Toulouse
Support Consultants
ATIE, M. Almaric, Z.I. La Plaine Basse, 81660 Route du Pont de l'arc
Fax: (33) 5 63 61 80 22
Drying Equipment
ALMATIFEL, PO Box 198, 84009 Avignon Cedex
Fax: (33) 04 90 14 87 30
Various Equipment

Italy

TURATTI, Viale Regino Margherita no. 42, 30014 Cavazere
Fax: (39) 0426 31 07 31
Drying Equipment
United Nations Food and Agricultural Organization (FAO), Via delle Terme
di Caracalla, 00100 Rome
Information Centre

Netherlands

BMA Nederland, Ampereweg 3/5, 3442 AB Woerdem
Fax: (31) 03 48 435 435
Various Equipment
KIT Royal Tropical Institute, Mauritskade 63, 1092 AD Amsterdam
Information Centre, Resource Centre

United Kingdom

Intermediate Technology Development Group (ITDG), Bourton Hall,
Bourton on Dunsmore, Rugby, CV23 9QZ
Tel: (44) 1926 634400
Fax: (44) 1926 634401, e-mail: infoserve@itdg.org.uk
Support Consultants
International Federation for Alternative Trade (IFAT), 30 Murdock Road,
Bicester, Oxon OX26 4RF
Tel: (44) 1869 249819
Fax: (44) 1869 246381, e-mail: info@ifat.org.uk
Information Centre
Natural Resources Institute (NRI), University of Greenwich, Central
Avenue, Chatham Maritime, Chatham, Kent ME4 4TB
Resource Centre

Appendix 15

GERES

Our mission

- Our team offers to associations, institutions, collectives and businesses, 20 years of experience in development engineering and a vast knowledge of problems facing countries of the South.
- Our programmes give priority to local resources within a context of lasting development that respects man and his environment.
- We operate in France, in countries in economic transition, and in developing countries.

Our team

- Initiates and manages development projects based on strong involvement of local operators, within a global approach: identifying needs, and putting in place institutions, appropriate technology and solutions.
- Helps owners to carry out their projects: feasibility studies, project set-up and technical follow-up.
- Carries out expert missions, feasibility studies, market surveys and applied research for development.
- Gives professional training and education to local operators on project management, quality management and new technologies.
- Capitalizes on its technical and methodological expertise.

Our operations

Developing economic operations

Supporting local traders in the food sector
- Accompanying groups of women and individual promoters working on fruit, vegetable and cereal processing for local markets (drying in Burkina Faso, various sectors in Burundi).
- Technical and commercial assistance for businesses in Burkina

Faso, Madagascar and Burundi working in the area of factory-drying of food products.

Support for local operators in the small-scale production sector
- Setting up a technical resources centre in Burkina Faso: accompanying businesses, studies, production and dissemination of information.
- Developing traditional ingredients for medicinal remedies in Burkina Faso.
- Improving quality of woven cashmere products in Ladakh (northern India).

Using solar energy for setting up income-generating activities
- Developing passive solar heating systems in the small-scale farming sector (glasshouses, chicken coops, production units) in Ladakh (northern India).

Energy demand-side management and promoting renewable energy

Domestic, manufacturing and service-sector energy demand-side management
- Study on demand-side management of electricity to improve electric supply networks in ACP region.
- Promoting demand-side management solutions (cold storage for industries, low-wattage lamps) in Morocco, Tunisia and France.
- Training hospital managers in Tunisia.

Promoting renewable energies
- Preparing investment projects in micro-hydro and wind-power fields in south of France.

Bioclimatic architecture
- Helping to construct bioclimatic buildings in Burkina Faso (schools, dispensaries, housing) and Tunisia (school, post office).

Local development in energy and environment fields

Domestic and community decentralized rural electrification
- Helping with decentralized rural electrification pilot programme in Morocco (90 villages).
- Setting up domestic and community electrification with photo-voltaic systems in Tanzania.

Increasing value of waste for energy and farming, waste processing
- Training and raising awareness of composting organic waste in local collectives and associations.
- Maximizing energy value of waste in wood sector.
- Developing decentralized solutions for reprocessing water.

Reducing consumption of firewood
- Drawing up socio-economic references for marketing stoves and domestic or small-scale production cookers with reduced firewood consumption in Cambodia.

Innovations and technology transfer

Design and implementation of equipment
- Drying equipment for domestic, small-scale and semi-industrial uses in Madagascar and Ivory Coast.
- Small-scale method for processing cashmere.
- Food storage warehouses.

Bibliography

Altersyal (1997) *Deux Décennies d'expériences sur les Alternatives Alimentaires* (*Two Decades of Experience in Alternative Food Products*), Charles Léopold Mayer, Paris.

Badini, Z. (1998) *L'approche des Marchés.* Documents de travail, session de formation (*Approaching Markets.* Working documents, training session), CEFOC/GERES, Ougadougou.

Bridier, M. and Michaïlof, S. (1995) *Guide pratique d'analyse de Projet* (*Practical Guide for Project Analysis*), 8th edn, Économica, Paris.

Broutin, C. (1997) Les produits céréaliers transformés secs: des nouveaux produits (Processed, dried cereal products: new products), *Bulletin TPA*, 97: 18–32.

Camilleri, J.L. (1996) *La petite Entreprise africaine: Mort ou Résurection* (*The African Small Business: Death or Rebirth*), l'Harmattan, Paris.

Geretsen, E. (1999) *Étude de Marché biologique européen des Produits séchés tropicaux*, rapport d'études (*Study of the European Organic Market for Tropical Fruits*, study report), GERES, Aubagne.

Hersan, C. (1987) *Pratique de l'Assurance de Qualité* (*Exercising Quality Control*), Collection Tec et Doc, Lavoisier, Paris.

Lopez, E. and Muchnick, J. (1997) *Petites Entreprises et grands Enjeux: Le Développement agroalimentaire local* (*Small Businesses and Big Challenges: Local Food Development*), l'Harmattan, Paris.

Outtier, A.C. and Traoré, A. (1998) *La Maîtrise de la Qualité en Séchage agroalimentaire*, Document de Travail, Session de Formation (*Controlling Quality in Food Drying*, Working Document, Training Session), CEFOC/GERES, Ouagadougou.

Pinot-Bernard, M. (1992) *Les Fruits séchés d'origine tropicale: le Marché* (*Dried Tropical Fruits: The Market*), Rapport COLEACP, Paris.

Rozis, J.F. (1995) *Sécher des Produits alimentaires* (*Drying Food Products*), Collection 'Le Point Sur', GERES, Aubagne.

Salles, I. (1994) *Étude du marché français de la mangue séchée* (*Study of French Market for Dried Mango*), Rapport d'études, GERES, Aubagne.

Thuillier, F. (1997) Choix technologiques (Technical choices), *Bulletin TPA*, 97: 24–43.

Treillon, R. (1992) *Innovations technologiques dans les Pays du Sud* (*New Technologies in Countries of the South*), l'Harmattan, Paris.

DRYING FOOD FOR PROFIT

A Guide for Small Businesses

Barrie Axtell

Practical, relevant, comprehensive – simple English throughout

Covers commercial, operational, business and technical aspects

Case studies

Key text for entrepreneurs and project overseers

This book is for existing and prospective entrepreneurs who wish to produce dry foods commercially at small and medium scale. Great effort has been made to use simple language but at the same time to examine all relevant technical aspects.

It starts with examining the basic principles of drying. This is followed by aspects related to markets, including advice on carrying out a market survey, and marketing or selling the product both locally, nationally and internationally.

On the assumption that a market exists, the publication then examines small-scale drying technologies, operational aspects related to the drying of common food groups and gives advice on establishing production, planning quality assurance and costing the product. The section ends with advice on preparing a business plan.

The final chapter considers the design of a dryer for a given application; technical calculations have been simplified so that those who can add, subtract, multiply and divide and calculate percentages will be able to design a dryer for any application. Case studies from Africa, Asia and Latin America are included.

Barrie Axtell is a food technologist and a small enterprise specialist who has spent 30 years working in Asia, Africa and Latin America

Published in collaboration with the Commonwealth Secretariat

ISBN: 1 85339 520 X
Price: £9.95
Publication Date: February 2002
Extent: 128pp
Format: 246 x 177 mm
Subject Category: Food Processing
Illustrations: Line drawings, photographs, index

*i***TDG**
PUBLISHING

ITDG Publishing, 103–105 Southampton Row, London, WC1B 4HL, UK
Tel: +44 (0)20 7436 9761 Fax: +44 (0)20 7436 2013 E-mail:
orders@itpubs.org.uk

www.ingramcontent.com/pod-product-compliance
Lightning Source LLC
Jackson TN
JSHW011352130125
77033JS00016B/570